Pop Music, Culture and Identity

Pop music lasts. A form all too often assumed to be transient, commercial and mass-cultural has proved itself durable, tenacious and continually evolving. As such, it has become a crucial component in defining various forms of identity (individual and collective) as influenced by nation, class, gender and historical period. Pop Music, Culture and Identity investigates how this enhanced status shapes the iconography of celebrity, provides an ever-expanding archive for generational memory and accelerates the impact of new technologies on performing, packaging and global marketing. The series gives particular emphasis to interdisciplinary approaches that go beyond musicology and seeks to validate the informed testimony of the fan alongside academic methodologies.

More information about this series at
http://www.palgrave.com/gp/series/14537

Laura Glitsos

Somatechnics and Popular Music in Digital Contexts

palgrave
macmillan

Laura Glitsos
Faculty of Humanities
Edith Cowan University
Perth, Australia

Pop Music, Culture and Identity
ISBN 978-3-030-18121-5 ISBN 978-3-030-18122-2 (eBook)
https://doi.org/10.1007/978-3-030-18122-2

Cover illustration: Danil Cetvericov / Alamy Stock Vector

This Palgrave Macmillan imprint is published by the registered company Springer Nature Switzerland AG.
The registered company address is: Gewerbestrasse 11, 6330 Cham, Switzerland

For Cain, and all the other music fiends

Acknowledgements

In some senses, this project has been 37 years in the making, as old as I am at the time of its publication. I say this because it is the culmination of all the impacts and experiences which have shaped my thinking and academic world up until this point. However, my academic mind is shaped not only by that which I have read and researched but also by the depth of substance from the unique people in my life. I have been lucky enough to receive the highest level of mentorship and support from the widest range of creative, inquiring and scholarly minds. I wish to thank these individuals with the deepest gratitude.

I thank my partner Cain Cressall, who continues to challenge and support me. This book would not be possible without his patience and wisdom.

I thank the Glitsos Family for their support in my academic career and for the continuing homecooked meals when I have been buried in writing. I would also like to thank the Cressall Family.

I thank my mentors. Professor Jon Stratton, who is always ready to reply to my endless emails and respond to my questions about the writing and publishing process. This work derives somewhat from my doctoral dissertation, which started with his support. I thank Professor Suvendrini Perera for her guidance in finishing my thesis and for introducing me to the field of somatechnics, from which I have gained so much.

I thank all my friends, who have provided indelible love and laughter over the years—you all know who you are. Trudging through sometimes laborious research work has been alleviated by these friendships. I also thank my friends for their patience when I have been locked away working on my research, and for welcoming me back with open arms when I re-emerge. I would especially like to thank my Mum, Patricia and my brother, Will.

I would like to make a very special acknowledgement to Tammy Geddes, whose courage and passion inspires me always. I would also like to make a very special acknowledgement to Anthony Roe, who never let me give up on myself and whose laughter helps me to face the world.

I would like to thank my colleagues: Mr James Hall, Dr Jessica Taylor, and Associate Professor Panizza Allmark at Edith Cowan University. Thank you to my colleagues at Curtin University, in particular, Dr Christina Chau, Dr Francis Russell, Mr Chris Mason, Dr Tama Leaver, Dr Eleanor Sandry and Dr Jo Jones. Thank you to Dr Madison Magladry for the help in the editing process. Thank you to all the supportive colleagues beyond, especially Dr Catherine Hoad, Associate Professor Andrew Hickey, Professor David Hesmondhalgh, and Professor Andy Bennett. A very special thank you to Dr Karyn Morrissey.

I wish to also thank Shaun Vigil, Glenn Ramirez, Camille Davies, and the whole team at Palgrave for their fantastic and professional expertise which has made the publishing process a pleasure.

Contents

About the Author

Laura Glitsos is a Senior Adjunct Research Fellow in the School of Arts and Humanities at Edith Cowan University in Perth, Western Australia. In addition to her expertise in the media and music industries, she has worked as a professional vocalist for which she was honoured with two Western Australian Music Awards. Dr Glitsos has published peer-review research in media studies, cultural studies, and popular music studies. *Somatechnics and Popular Music in Digital Contexts* is her first sole-authored book. She continues to lecture in media and cultural studies in Western Australia.

CHAPTER 1

Introduction: Listening Through the Body

There is more sagacity in thy body than in thy best wisdom.
Nietzsche ([1883] 2009, 16)

Introductory Thoughts

Following the epigraph from Nietzsche above, this work is both a celebration and explication of the body in the world and the ways that our body situates our consciousness as a lived formation, one which is oriented by the experience of music listening ([1883] 2009). However, philosophical traditions may never succeed in apprehending the extent of bodily wisdom, largely because apprehending it thus conflates and reduces it. Gayatri Spivak refers to the body as that which "cannot be thought" (1994, 177). The body may be *biologically* thought-through, perhaps, but Spivak would say biology is only "one way of thinking the systematicity of the body" (177). Rather, she ruminates that:

> If one really thinks of the body, there is no possible outline of the body as such … There are thinkings of the systematicity of the body, there are value codings of the body. The body, as such, cannot be thought, and I certainly cannot approach it. (Spivak 1994, 177)

L. Glitsos, *Somatechnics and Popular Music in Digital Contexts*, Pop Music, Culture and Identity,
https://doi.org/10.1007/978-3-030-18122-2_1

From the very start, this book begins with both concession and paradox, in that this work pursues an examination of that which cannot be entirely known. Yet, I would marvel that this is the compelling status of the body in the world as it is torn between—and sewn together with—language, materiality, technology and philosophical treatments, which are all, ultimately, technicities. Though we might never fully reach a complete 'outline' of the body, we might still lend ourselves to a project which touches, perhaps even only briefly, on the breathtaking potentialities of its scope.

This book, of course, is not only about the body, but how an individual mobilizes bodily systems through and with technicities for the pursuit of listening pleasure in popular music. In order to approach such a task, this book borrows tools and language from a range of fields, in particular, from Cultural Studies, Popular Music Studies, Philosophy, Somatechnics, and Internet Studies. Thus, this study is multidisciplinary in its approach in that it crosses a range of modes in the category of social inquiry. I would briefly note, too, that this book focuses on the function of music *listening*, as opposed to the experience of making music or producing music. However, at times, I may refer to the creation and production of music, but only as it relates to constructing the experience for the listener, as these processes often converge (which we will uncover especially in the discussion of live music).

Moreover, this book situates the listening subject in the context of contemporary digital culture. How can we think through the digital? Technically, it is a process whereby data is translated and stored in binary form as zeros and ones. This is a process which emerges from an ancient means of counting and mathematics, that is, a mechanism (or technicity) through which to interact with the world of numbers (Peters 2016). Benjamin Peters contemplates the historical narrative of 'the digital discourse,' suggesting it dates back to "the Latin source of the term itself—the original digit, or the index finger" (94). He writes that:

> Digits do what index fingers do—namely, count, point, and manipulate. … Ever since we evolved *extensor digitorum* muscles, ours has quite literally been what media theorist Teil Heilmann calls a "digital condition": *digital media do what fingers do.* (2016, 94; my emphasis)

This remark is a prescient summary of the following research, considering my interest here is in the production of digitization and the body as they emerge as simultaneous artifacts, the primary conditions of which are perhaps not so far apart. In fact, Nick Prior suggests that if "we stripped the digital back down to the digit" we would be "reuniting fingers and technologies to the origins of counting" (2018, 123). In a striking parallel, he also posits that this resonates with the examination of popular music and digitization, because, "Maths and music are close bedfellows, as we know" (123). In relation to somatechnics, too, I would suggest that this conceptual pairing—that is, the relationship between the finger (sôma) and digitization (techné)—positions the arguments in this book as primary entanglements from the very beginning. The digital, in its ancient roots, is primarily corporeal.

In moving forward to the contemporary world, I put forth digital culture as a formation of discourses, and practices that revolve around an increasing reliance on complex computerized systems. We might think of 'formations' following Prior, as "loose configurations, less rigid than institutions but characterized by constituent material and non-material elements sharing enough properties in common to produce systemic effects" (2018, 14). Prior's characterization is vital to the context of this project because I suggest that the 'digital music context' should be thought about with just as much emphasis on new technologies and new media devices as the production of meaning around the value and affective potential of listening practices.

To expand on this point, I suggest that new media forms do not cultivate new emotional experiences *only* for individuals who are active in digital Internet technology but for everyone situated within the framework of the digital context. It is the social and historical moment of digitization as a contemporary phenomenon that I theorize, rather than just those practices we can single out as specifically digital (although, I do take single practices as examples in order to illustrate my points). For instance, listening to music on the radio in a contemporary context implies a very different experience than what it meant for individuals listening to music on the radio in the 1920s. The materiality of the practice is mostly the same (i.e., the radio unit is still a wireless device that transmits audio content programmed by someone other than the user). However, the affective encounter is different because, in the 1920s, listening to music on the radio was brand new; a "magical" phenomenon where songs could be snatched out of "thin air" and were free to anyone (Fisher 1926, 12). Radio meant the liberation of music and content. It also led to the emergence of new family bonding activities because the family unit could sit and listen to the radio together, singing songs and listening to favorite programs.

Now, radio is considered a secondary medium that many people use for background noise or while driving in their car (Berland 1990, 179). The way the listener is affected by music as it is mediated through the radio has changed in both meaning and intensity and continues to change against shifting historical contexts. In Althusserian terms, the subject is 'always-already' presupposed in and by the cultural lexicon (1971). This is to say that digitization is working upon the listening experiences of everybody who is situated within the paradigm of digital technology, which is now a global phenomenon.

Of course, while digital culture is global, it is by no means universal. Digital culture crosses many geographic and geopolitical boundaries but still excludes participants along demographic lines. Many people do not have access to digital or Internet technologies as a result of various factors such as financial inequity or Internet censorship restrictions. This inequality is referred to as the digital divide (Castells 2001; Kirkman et al. 2002; Norris 2001). Therefore, I situate this argument in what are the privileged communities participating in digital culture. Levels of participation vary not only across nations but within nations, such as the disparity which is often found between rural and metropolitan regions. Digital culture cannot be considered as an exclusively Western phenomenon either; many non-Western nations have digital and Internet technology of varying speeds and accessibility. However, it is important to note that the uptake of, and access to, Internet technologies is considerably less in developing nations (Guillén and Suárez 2005, 681). This imbalance does result in disenfranchisement for many people living in these communities, particularly concerning education or employment opportunities and should be noted in any study of 'global' digital culture.

Somatechnics as Theoretical Tool and Mode of Apprehension

While situated in the context of digitization, I pursue this examination using the theoretical tools of somatechnics. Somatechnics calls for a recognition of the body in the world as an artifact wrapped up, entangled, and *produced* by the materialities of that world (Sullivan and Murray 2009). The field emerges from a series of conferences and research collaborations based at Macquarie University in Sydney involving Nicki Sullivan, Joseph Pugliese, Samantha Murray, and Elizabeth Stephens, as well as Susan Stryker from the U.S. (Pugliese and Stryker 2009, 1), who wished to collapse the counterfeit binaries delimiting the discourse of human-technology relationships. It is also critical to note that since its emergence, the field of somatechnics has attended to questions of race. The first issue of the *Somatechnics* journal was edited by Suvendrini Perera and Joseph Pugliese, titled "Combat Breathing" (2011) and was concerned with the effects of transnational state violence on the body.

Somatechnics puts forth, in its very etymology, "the inextricability of sôma and techné" (2012, 3). That is, the term emphasizes that soma, or 'the body,' cannot be thought about other than in its production by and through *in-worldness*. Thus, it is an approach to understanding the body as irretrievable and indistinguishable from that which produces that body in the world, both through language and the technics of being and becoming. Sullivan explains that somatechnics, as a field, can be thought about in terms of:

> The bodily-being-in-the world, and the *dispositifs* in and through which corporealities, identities, and difference(s) are formed and transformed, come to matter … Somatechnics, then, supplants the logic of the 'and' (thereby moving beyond instrumentalist logic), suggesting that technés are not something that are added or applied to 'the body,' nor are they simply tools the already-constituted body-subject manipulates to its own ends. Rather, technés—in the Heideggerian sense—are techniques and/or orientations (ways of seeing, knowing, feeling, moving, being, acting and so on) which are learned within a particular tradition or ontological context (are, in other words, situated), and function (often tacitly) to craft (un)becoming-with in very specific ways. (Sullivan 2012, 4)

Heidegger casts a long shadow over somatechnics, and this particular work, as well as other phenomenologists such as Merleau-Ponty, for whom technics and technicity were recognized as the constituting principle of the corporeal status.

To understand the human relationship to technology is first to understand technics. In his exploration of Stiegler, Daniel Ross begs the question that, "If technology is the discourse on technics, what is technics itself?" (2017, 9). To which, Ross suggests that technics refers to "all the domains of skill, including not just cooking or dance but, for example, politeness, elegance, or poetry, even language itself. All human action has something to do with *tekné*" (9). This assertion is fundamental in that understanding human experience can never be untied from understanding how the human body maintains itself in the world. In effect, all human action is produced in some way by the technical aspects of the material world. Technical aspects may range from the most complex of technologies, such as a smartphone, to the acquisition of knowledges in order to hone techniques for survival, such as hunting or gathering skills. Many skills, too, require some technical object to support these interactions with the world, even something as simple as a rock can be useful as technicity. Daniel Ross takes Stiegler's apprehension of these kinds of objects to explain that:

> Technical objects are inorganic organized beings, possessing their own dynamic, irreducible to either physics or biology. Such inorganic organized beings are constitutive of both temporality and spatiality, these being the derivative decompositions of speed. If life is the conquest of mobility, technics, as a process of exteriorization, is the pursuit of life by means other than life. (Ross 2017, 3)

Comprehending the fundamental nature of technics is critical in understanding the basic premise of somatechnics, and the following book traverses the widest variety of technologies and techniques—from iPods through to dancing skills—to think about their roles in producing the subject in the material world.

The argument of this book is then twofold: In the first instance, it is arguing for the use of somatechnics in popular music studies. In the second instance, it is arguing that the shifts in listening experience for popular music fans brought about by digital configurations can be articulated through the somatechnics model in order to understand that it is not just the technologies that are changing, but that the *historical moment of the bodymind* is changing with them interactively. In pointing to the music-bodymind relationship as 'interactive,' I would mandate from the outset a kind of affective phenomenal epistemological approach, thoroughly in line with the language of somatechnics, and one that perhaps poses affect as its basis, as somatechnics largely does. Music affects the bodymind, and in some ways, through that encounter, music becomes an aspect of the bodymind, that is, not separate but woven through and absorbed as *tekné* of the somatic, its fleshy substrate.

Bodymind: Thinking About the Body

The configurations of the body, or what I will often be referring to as the 'bodymind,' not only provide us with our ontological framework but also the tools with which to challenge, fragment, and for brief moments, perhaps even fracture that ontological framework. What I mean, in one respect, is that for one to challenge the conceptualization of the body or a 'bodymind,' one must first speak through a bodymind, that is, in the oldest philosophical tradition, consciousness may only speak on behalf of, and within, its moment of consciousness (Merleau-Ponty [1945] 2013, 74). The bodymind then calls forth its potential through the very recognition of its limitations.

What is it to speak of the bodymind—and why am I using this term as opposed to 'the body'? To explain this, and to seat this term theoretically for the remainder of the book, I borrow from the Lacanian model of subjectivity that Elizabeth Grosz articulates in *Volatile Bodies*: that of the Möbius Strip (1994, xii). The Möbius surface is defined as having only one side and only one boundary component. A model of the Möbius can be made by taking a paper strip, giving it a half twist, and joining the strip together. A line can be drawn from a starting point, in a single continuous curve through the length of the strip, but without ever crossing an edge. As Grosz explains, this model:

> avoid[s] many of the common metaphors that have been used to describe the interactions of mind and body, metaphors of embodiment, of containment, machine metaphors, two-sided coins, hydraulic models—models which remain committed to dualism [...] The Möbius Strip has the advantage of showing the inflection of mind into body and body into mind, the ways in which, through a kind of twisting or inversion, one side becomes another. (1994, xii)

Using the Möbius metaphor is not to suggest that I do not and will not use the separate terms 'body' and 'mind,' although when I do, it is to deliberately differentiate when we might indicate a process as being more heavily weighted in the mind or in the body. It is not to negate the bodymind configuration, but to note its complexity and continuum. For example, the mental formations which organize thoughts and become complex systems of interactions with the world through language and identification schemas are produced in the mind. However, they are not separate from the body in that consciousness, as well as all the mental formations converging and diverging in thought processes which emerge from that conscious being, is embedded within, and is contingent upon, on the organic and material facticity of the body. In addition to this, the embodied experience gives rise to how one is produced in the world and what shape those thoughts might take particularly along racial and gendered lines.

One aspect of the bodymind simply does not exist in clear terms without the other, except perhaps in cases in which an individual suffers extreme brain trauma and is left with only some autonomic processes which continue to animate the organic aspect. In these cases, the absence of mind only problematizes the definitions of human experience, with debate emerging around whether or not that individual is 'truly alive' and whether or not that individual's body should be left to perish. Ethical conundrums aside, I consider that the music listening experience is a process that calls upon the bodymind configuration in its totality. The following book then explores the various ways that this happens.

Coupled with the practice of music listening, we may consider how the boundaries and borders of the bodymind are joyfully circumnavigated, particularly with the introduction of technologies which interrogate or interrupt those practices. However, we may also consider how certain boundaries and borders are also radically reshaped (for example, in the new technologies coupling iPods with vibrators). In other ways, the vernacular use of some technologies, such as the iPod or even a set of headphones, lull the listener into a mundane set of expectations and produce an 'everydayness' to the relationship between music and bodies. At times, I seek to denaturalize these practices to produce a theoretical moment of consideration. Other times, I may point to as-yet unrealized opportunities of the music-body interface as possibilities for future undoing or redoing of bodily interactions.

As I have started to reveal, in this work, the bodymind system is not some separate artifact operating distinctly from the world but forms part of the wider, complex architecture in which it is entrenched. However, perhaps paradoxically, this is not to suggest an *a priori* understanding of the body in fixed terms but to concede that, as an object of culture, the body cannot be taken outside of its relationships with the techniques and organizing practices of the material world, particularly because the body is expressed by the language of those organizing practices. In fact, far from suggesting any *a priori* bodily constructs, as Michael O'Rourke and Noreen Giffney offer, "Somatechnologies harbor possibilities for disruptions, counter-actualizations, destabilizations, and for the creation of new selves, affinities, kinship relations, and cultural possibilities" (as cited in Sullivan and Murray 2009, x). How wondrous that all these possibilities and potentials might be viewed, theorized, and conceived in popular music and its study. That is then the purpose of this project.

Why Popular Music and Somatechnics?

Up until now, somatechnics has largely (although not exclusively) been concerned with "queering the technologization of bodies" as a way to explore the political implications of bodies in culture (Sullivan and Murray 2009). However, deployment of somatechnics in more diverse fields is beginning to expose the use of its scope (Pugliese and Stryker 2009, 2). At its core, however, somatechnics seeks to expose the invisible nature of the way technicities produce and are produced by the bodymind. Thus, the integration of popular music studies with somatechnics *presupposes* a political aspect to this project. This is to say that the ideas presented in this work are underwritten by the political infrastructure of somatechnics.

Further, both somatechnics and popular music are born from cultural studies, and it is perhaps for this reason that both fields are concerned foremost with the political implications of the lived form. Popular music studies' history with the political cannot be overstated and thus its articulation through the cultural studies is critical. Gilbert Rodman sees a natural accord between popular music and cultural studies because popular music is a site "where significant cultural and political struggles manifest themselves" (2015, 49). As Rodman points out, the legendary theorist Stuart Hall noted this crucial nexus by stating that:

> Popular culture is one of the sites where this struggle for and against a culture of the powerful is engaged: it is also the stake to be won or lost in that struggle. It is the arena of consent and resistance. It is partly where hegemony arises, and where it is secured. It is not a sphere where socialism, a socialist culture—already fully formed—might be simply 'expressed.' (Hall 1981, 232 as cited in Rodman 2015, 49)

Therefore, the kinds of practices performed in and through popular phenomena, such as contemporary music, both expose and form some of the most compelling sites of power and rebellion. The notable popular music scholar Simon Frith (who some may even call the 'grandfather' of popular music studies) emphasized how crucially the matrices of power and resistance play out in the identity-making mechanisms provided through interconnections with popular music practices, noting "In 'possessing' music, we make it part of our own identity and build it into our sense of ourselves" (Frith 1987, 143 as cited in Rodman 2015, 55). Andy Bennett and Steve Waksman explain the ways in which popular music studies built on from the "scholarly blueprint offered by Stuart Hall, Dick Hebdige, Angela McRobbie, Paul Gilroy and others who taught or studied at the Birmingham Centre for Contemporary Cultural Studies" (2015, 3). Bennett and Waksman call to mind the foundations of popular music, which developed from the late 1960s and which were

> punctuated at intervals by the publication of highly influential books such as Dave Laing's (1969) *The Sound of Our Time*, R. Serge Denisoff and Richard A. Peterson's (eds.) (1972) *The Sounds of Social Change*, Wilfred Mellers's (1973) *Twilight of the Gods* and Simon Frith's (1978) *The Sociology of Rock* (republished in 1981 as *Sound Effects*). As the scholars associated with these titles collectively illustrate, from its very beginnings, the academic study of popular music was a multi-disciplinary affair, and this is something that has remained a centrally defining feature. (Bennett and Waksman 2015, 1)

Since its inception, the academic study of popular music has been highly critical and multi-dimensional, and it is perhaps for this reason why the field of somatechnics and popular music studies resonate so beautifully together.

Sticky Listening: How to Talk About Pleasure and Listening

In one sense, I like to think of that which is pleasurable as that which 'sticks.' For instance, when we love a song, it 'sticks' to us. It plays over in our mind, and we might replay it in a hum during the course of a day. I take "stickiness" (Ahmed 2004) as the launching point for an introductory discussion on the relationship between somatechnics, affect and music. Stickiness, as an affective and political encounter, can be theorized as exemplary of the somatechnics approach because it conceptualizes the contacts, imbrications, and absorptions between *soma* and *tekné*, the organic and the inorganic, the fleshy and the mechanical.

To think through music and affect using the language of 'stickiness' points to Melissa Gregg and Gregory Seigworth's position that affects are "resonances that circulate about, between and sometimes *stick* to bodies and worlds, and in the very passages or variations between these intensities and resonances themselves" (2010, 1; my emphasis). Music, though a non-human entity, is a force that can quite literally resonate with the human body—with a heartbeat or a breathing pattern for example. Music can travel through the body, stick to the body and circulate around the body in ways that seem to cross intimate borders. Music penetrates the ear and also moves through the hair; we cannot see music (in its 'pure' form as a sonic event in which vibrations move air), but it nonetheless shrouds the body and brings with it the specters of its discursive, language-bound mechanisms. For Sara Ahmed, the theoretical language of stickiness also serves to explicate the encounters between bodies, surfaces, and even sites of meaning-making. In Ahmed's thought, affect is the very thing that demarcates where one body ends, and another begins. In terms of stickiness then, affect is the moment of "contact with others," or how the Other sticks to the self (2003, 194). In one way, music listening is only defined by its technological mediation, and certainly, this becomes the predominant issue in a context of digitization where mediation technologies have become many and varied. The stickiness of music can be read in the sense of mediation in a variety of ways, for example, the stickiness of vinyl, the stickiness of bodies-to-bodies at live music shows, the stickiness of dirty fingers on greasy touchscreens—all of which I shall be exploring in the following chapters.

Music 'stays' with us, shapes our emotional worlds and moves with us through our day. Music is so personal to the body but is not the body, yet at times seems to *become an aspect of that body*, as would a contact lens or even further, food once separate, then taken in to become digested material, and finally part of the blood itself. It is utterly collapsed into the body. In a similar sense, the line between where the bodymind starts and aural phenomena ends is blurred and finally, sometimes, utterly erased. This relationship between the human body and music can, therefore, encompass the widest possible range of affective phenomena, which I take as a foundational epistemological stance.

Chapter Outline

In the following chapter, "Materiality, The Bodymind, and Music Listening," I explore the ways in which sensory perception comes to render music as both knowable and pleasurable through the interplay of bodymind with materiality—that is, through the complementary interactions with surface, weight, touch, feel, smell, size, and so forth. I will argue that the changing notions around corporeality, knowability, perception, and materialism come to reshape our relationships to the physical objects of music, which shapes listening pleasure more broadly. To do this, I traverse several of the sensory mechanisms including; the visual system, the auditory system, the olfactory system, the tactile system, and the proprioceptive system.

In this chapter, I especially emphasize that the role of touch and materiality maintain a critical, yet redefined, position in music listening culture despite the emergence of digital music as the dominant listening mode. This will follow Dominik Bartmanski and Ian Woodward's assertion from their work on the vinyl record, in which they suggest that:

> Recent discussions of music listening practices have given priority to the digitalization of sound and the role of digital music players in changing the form, medium and possibly even the content of listening. While such an emphasis is warranted given the rapid uptake of digital music consumption, it is also the case that vinyl records are currently the fastest growing area of music sales. (2015, 3)

The authors point out a critical context for this work in that they note that despite the radical and fast-moving shifts in music consumption and commodification, material music practices such as the compact disc and vinyl record, still maintain a role in the pleasure of listening, for various reasons. I follow Bartmanski and Woodward's argument up to a point, and I ruminate on similar themes such as ritual and aura (2015, 7). However, I branch off from this approach in order to relocate the argument within a different framework that approaches the field using a concept which I term as sensorial somatechnics—a concept which directly follows Nikki Sullivan's "somatechnics of perception" (2012). Using this model gives us the opportunity to better understand our affective encounters with music listening in the transition between material and immaterial music products. This is also to acknowledge the way in which historically specific listening practices that emerged with material practices helped to cultivate the discursive structures that are in transition today, and why they retain significance. By using this basis, I will assert that the significance of touch, and other sensorial mechanisms such as smell, is still very much a critical aspect of listening in the digital landscape, because of long-entrenched relationships between value and materiality.

The following chapter, Chap. 3, focuses on liveness in the age of digitization. This chapter elucidates the various ways in which we might imagine new distinctions between liveness and mediatization in the post-millennium digital context by applying theoretical perspectives of somatechnics to certain live music listening practices. I argue that there are certain technics which bring forth and produce listening pleasure, functioning with and within the somatic. Namely, I focus on: The technics of time, and its significance in the production of liveness and authenticity; the somatechnics of dance as it functions to reinterpret music listening, bodily and psycho-spatial boundaries; the somatechnics of vocality as it functions in the live music space as a vibratory technology; and finally, through the use of Derrida's work on 'teletechnology' I examine the emergence of the camera phone technology in reshaping, recrafting and reimagining new understandings of liveness, presence, and actuality in the realm of the concert experience. The various technicities explored in the chapter are not static, rigidly fixed processes which produce the relationships between pleasure, liveness, and music. Rather, we see the way the relationship between these crucial aspects are constantly in flux and thus continue to change both our understandings and experiences of 'listening pleasure' in the live context.

Chapter 4 is titled "Screen as Skin: The Somatechnics of Touchscreen Music Media" and explores the way mobile music devices with touchscreen technology produce new 'somatechnical' figurations that reshape the emotional dynamics of music listening. Touchscreens imply the relationship between skin-on-skin—the skin of our body (in particular the hands) against the skin of the screen. It follows that mobile touchscreen devices suggest a degree of sensuality—in the coming together of bodies, fluids and other organic materials which 'stick' to the touchscreen, the language of 'stickiness' pointing again to Ahmed's conceptualization of the way affect can "stick" to bodies which I discussed earlier. The function of skin, both in a corporeal and a discursive sense, cannot be overstated. Skin is the covering that "protects us from others and exposes us to them" (Cataldi 1993, 145). Skin is profoundly significant in that it provides the basis for an overwhelming variety of trends in the politics of subjectivity, from the classic work of Frantz Fanon on skin colour (see *Black Skin, White Masks* [1952] 1967) to more recent feminist work on the politics of ageing skin or stretching skin, such as in the case of pregnancy (Tyler 2003). Skin is not politically benign. It is "the fleshy interface between bodies and worlds" (Ahmed and Stacey 2003, 7). By "thinking through the skin," to use Ahmed and Stacey's words, I read mobile touchscreen technology as an exciting new way to imagine music listening in terms of cyborgian relations.

Chapter 5 focuses on the Internet as a technology of music listening which is accessed exclusively through fixed-point personal computers. I am concerned with how the listener actively and creatively produces their own listening experience as it is mediated through this particular configuration. This line of thought builds somewhat on the work of David Hargreaves, Jonathan Hargreaves, and Adrian North in which they expound on the significance of "imagination and creativity in music listening" and through which they understand "perception as creative construction of knowledge" (2011, 156). I take this further to argue that the coupling of the bodymind and the personal computer produces a unified field of relations, an artifact which we can examine through the lens of somatechnics. This unified field of relations, what I will refer to from now as the 'personal computer bodymind,' or for short the PCBM, realigns and reconstitutes listening pleasure in unexpected ways. The PCBM configuration enables unique and creative listening practices contingent upon the functional nature of Internet-computing technologies as they are accessed in Web 2.0 culture. I argue here that the PCBM produces a new 'type' of listener: the creative listener.

In the final chapter, I reflect on several practices of music listening which have emerged in the past decade and which integrate virtual or mixed realities such as TheWaveVR and 3D mapping at live shows. By doing so, I wish to put forth an understanding of where the emphasis of music listening is located and in which directions it may continue. What I propose is that we see an emerging trend toward virtual and mixed reality technologies in music listening which play upon notions of 'the real' (and which may or may not continue as a lasting preoccupation in one form or another). Most importantly, I expound on what this might tell us about the bodymind in the music listening experience and how this experience continues to be produced by techné. Building from Mark Hansen's work in *Bodies in Code* (2006), I offer that next-generation virtual reality and 'mixed reality' technologies reconstitute the somatechnic capacities of bodymind dynamics in listening practices in a way that foregrounds the site of the body as the 'ground zero' of all experiential exchange. More simply, these new practices indicate that what one hears, sees and does in space ceases to matter as much as the *sense* of hearing, seeing and doing, through the complex system of the bodymind.

Concluding Remarks

To end, I wish to emphasize that this book is not about proselytizing shifts in technologies. This work is about recognizing the shifts in the bodymind as they relate to the technicities which produce them, and thus come to reshape notions of listening pleasure. In this regard, the central focus of this research is really 'the embodied subject,' or what Penelope Gouk and Helen Hills might call "the feeling subject" (2005, 19). I am concerned with exploring how 'the feeling subject' imbricates with the social, political, technological and economic dimensions of culture, dimensions in ways that are constantly in flux.

There is not just one way to sense our place in the world. Rather, there are innumerous and manifold "ways of feeling" (see Glitsos 2016) that suggest various states of intensity which are always in negotiation with both inner and external worlds. Not that the inner and outer worlds are distinct entities either. The inner and outer dimensions of being are always interacting with each other and then folding back on themselves in repeating patterns of echo and reverberation, in suggestion of Grosz's Möbius model. This book offers some new ways of thinking through those echoes and reverberations.

References

Ahmed, Sara. 2004. *The Cultural Politics of Emotion*. Edinburgh, Scotland: Edinburgh University Press.

Ahmed, Sara, and Jackie Stacey. 2003. *Thinking Through the Skin*. Hoboken, NJ: Taylor & Francis.

Althusser, Louis. 1971. *Lenin and Philosophy, and Other Essays*. New York: Monthly Review Press.

Bartmanski, Dominik, and Ian Woodward. 2015. The Vinyl: The Analogue Medium in the Age of Digital Reproduction. *Journal of Consumer Culture* 15 (1): 3–27.

Bennett, Andy, and Steve Waksman, eds. 2015. *The SAGE Handbook of Popular Music*. London: SAGE.

Berland, Jody. 1990. Radio Space and Industrial Time: Music Formats, Local Narratives, and Technological Mediation. *Popular Music* 9 (1): 179–192.

Castells, Manuel. 2001. *The Internet Galaxy: Reflections on the Internet, Business, and Society*. Oxford: Oxford University Press.

Cataldi, Sue. 1993. *Emotion, Depth, and Flesh: A Study of Sensitive Space, Reflections on Merleau-Ponty's Philosophy of Embodiment*. Albany: State University of New York Press.

Fanon, Frantz. [1952] 1967. *Black Skin, White Masks*. New York: Grove Press.

Fisher, William Arms. 1926. The Radio and Music. *Music Supervisors' Journal* 12 (3): 8–72.

Glitsos, Laura. 2016. Ways of Feeling: The Transformation of Emotional Experience in Music Listening in the Context of Digitisation. PhD diss., Curtin University.

Gouk, Penelope, and Helen Hills. 2005. *Representing Emotions: New Connections in the Histories of Art, Music, and Medicine*. Aldershot: Ashgate.

Gregg, Melissa, and Gregory Seigworth. 2010. *The Affect Theory Reader*. Durham, NC: Duke University Press.

Grosz, Elizabeth. 1994. *Volatile Bodies: Toward a Corporeal Feminism*. St. Leonards, NSW: Allen and Unwin.

Guillén, Mauro F., and Sandra L. Suárez. 2005. Explaining the Global Digital Divide: Economic, Political and Sociological Drivers of Cross-National Internet Use. *Social Forces* 84 (2): 681–708.

Hansen, Mark. 2006. *Bodies in Code*. New York: Routledge.

Hargreaves, David, Jonathan Hargreaves, and Adrian North. 2011. Imagination and Creativity in Music Listening. In *Musical Imaginations: Multidisciplinary Perspectives on Creativity, Performance and Perception*, ed. David Hargreaves, Dorothy Miell, and Raymond MacDonald, 156–172. Oxford: Oxford University Press.

Kirkman, Geoffrey, Peter Cornelius, Jeffrey Sachs, and Klaus Schwab. 2002. *The Global Information Technology Report 2001–2002: Readiness for the Networked World*. Oxford: Oxford University Press.

Merleau-Ponty, Maurice. [1945] 2013. *Phenomenology of Perception*. Hoboken, NJ: Taylor & Francis.

Nietzsche, Friedrich. [1883] 2009. *Thus Spake Zarathustra*. London: Dent.

Norris, Pippa. 2001. *Digital Divide: Civic Engagement, Information Poverty, and the Internet Worldwide*. Cambridge: Cambridge University Press.

Peters, Benjamin, ed. 2016. *Digital Keywords: A Vocabulary of Information Society and Culture*. Princeton, NJ: Princeton University Press.

Prior, Nick. 2018. *Popular Music, Digital Technology and Society*. London: Sage.

Pugliese, Joseph, and Susan Stryker. 2009. The Somatechnics of Race and Whiteness. *Social Semiotics* 19 (1): 1–8.

Rodman, Gilbert. 2015. Waiting for the Great Leap Forwards: Mixing Pop, Politics and Cultural Studies. In *The SAGE Handbook of Popular Music*, ed. Andy Bennett and Steve Waksman, 48–63. London: SAGE.

Ross, Daniel. 2017. A Summary of Bernard Stiegler, Technics and Time, 1: The Fault of Epimetheus. Academia.edu/DanRoss.

Spivak, Gayatri. 1994. In a word. In *The Essential Difference*, ed. Schor Naomi and Elizabeth Weed, 151–184. Bloomington: Indiana University Press.

Sullivan, Nikki. 2012. The Somatechnics of Perception and the Matter of the Non/Human: A Critical Response to the New Materialism. *European Journal of Women's Studies* 19 (3): 299–313.

Sullivan, Nikki, and Samantha Murray, eds. 2009. *Somatechnics: Queering the Technologisation of Bodies*. Surrey: Ashgate.

Suvendrini, Perera, and Joseph Pugliese. 2011. Introduction: Combat Breathing: State Violence and the Body in Question. *Somatechnics* 1 (1): 1–14.

Tyler, Imogen. 2003. Skin Tight: Celebrity, Pregnancy, and Subjectivity. In *Thinking Through the Skin*, ed. Sara Ahmed and Jackie Stacey, 69–83. Hoboken, NJ: Taylor & Francis.

CHAPTER 2

Materiality, the Bodymind, and Music Listening

Introduction

In this chapter, I explore the ways in which sensory perception comes to render music as both knowable and pleasurable through the interplay of bodymind with materiality—that is, through the complementary interactions with surface, weight, touch, feel, smell, size, and so forth. I will argue that the changing notions around corporeality, knowability, perception, and materialism come to reshape our relationships to the physical objects of music, and I will elucidate how this shapes listening pleasure more broadly. To do this, I traverse several of the sensory mechanisms including; the visual system, the auditory system, the olfactory system, the tactile system, and the proprioceptive system.[1]

In this chapter, I especially emphasize that the role of touch and materiality maintains a critical, yet redefined, position in music listening culture despite the emergence of digital music as the dominant listening mode. This will follow Dominik Bartmanski and Ian Woodward's assertion from their work on the vinyl record, in which they suggest that

> Recent discussions of music listening practices have given priority to the digitalisation of sound and the role of digital music players in changing the form, medium, and possibly even the content of listening. While such an

[1] I exclude gustatory, interoceptive and vestibular sensory systems.

L. Glitsos, *Somatechnics and Popular Music in Digital Contexts*, Pop Music, Culture and Identity,
https://doi.org/10.1007/978-3-030-18122-2_2

> emphasis is warranted, given the rapid uptake of digital music consumption, it is also the case that vinyl records are currently the fastest growing area of music sales. (2015, 3)

I follow Bartmanski and Woodward's argument up to a point, and I ruminate on similar themes such as ritual and aura (2015, 7). However, I branch off from this approach in order to relocate the argument within a different framework that approaches the field using a concept which I term as sensorial somatechnics—a concept which directly follows Nikki Sullivan's "somatechnics of perception" (2012). I will elucidate further on this shortly. For the moment, I will state that this line of thinking imagines the body's sensory mechanisms as techniques and technology that organize the affects and encounters of music listening. Simultaneously, as well as congruously, cognitive faculties organize meaning based on discursively produced constructs dependent on an individual's subjectivity and cultural positionality. Through deploying this theoretical perspective, I redirect the conversation from themes such as iconicity, the sensibility of coolness, and nostalgia, which have been well-executed elsewhere (Bartmanski and Woodward 2015; Green 2017) towards an exploration of the way in which we can read listening pleasure and materiality in the context of the somatechnics of perception, or 'sensorial somatechnics'.

For the purposes of this argument, I define material listening practices as any activity in which the listener deploys a material product on which music is inscribed. This includes vinyl records, cassette tapes, compact discs, VHS, and beta disc. I define immaterial listening practices as activities in which one listens to music which is not inscribed on a material object, and is only deployed in the MP3, WAV or digital format. However, we encounter the first paradox (of which many are to follow in this book) in which one must recognize that in fact 'immaterial' practices are nonetheless contingent upon material media; a computer, an iPod, an iPad and other mediators. So even 'immaterial' practices nevertheless call forth a technicity, an exteriorization. Even streaming services must be installed on a material device, through which the listener develops some physical connection. Thus, I refer to 'digital music' or 'immaterial music' as the historical context of post-millennium listening practices rather than as a descriptor regarding the facticity of the nature of materiality itself.

Sensorial Somatechnics

The project of this chapter is to cultivate a 'sensorial somatechnics' which borrows directly from Nikki Sullivan's notion of a "somatechnics of perception" (2012) in order to read listening pleasure as relational to materiality. Following Linda Alcoff and Merleau-Ponty, Sullivan asks us to consider perception as "both the vehicle and effect of a particular situated somatechnics, an orientation to the world in which the I/eye is always-already co-implicated, co-indebted, co-responsible" (302). For Sullivan, the relationship between faculties of apprehension and the production of meaning are inseparable, an idea which fertilizes the ground of somatechnics itself. As the reader will recall was mentioned in the Introductory Chapter, the term 'somatechnics' was coined by a group of academics "in order to highlight ... the inextricability of soma and techné, of bodily-being-in-the-world, and the *dispositifs* in and through which corporealities, identities and difference(s) are formed and transformed, come to matter" (302). Using this approach, Sullivan deploys a basis, what she calls "white optics," in order to unpack critical race relations and demonstrate the ways whiteness functions as a perceptual schematic mode, that is, a way of seeing (303). For example, Sullivan recounts an anecdote from an eighteenth century journal written by Janet Schaw, a 'Lady of Quality', who visited Antigua in 1774, which "tells us much about the somatechnics of perception and its role in the somaticization, the coming-to-matter" of things (302).

> Proceeding on foot from St John's Harbour to her lodgings, the newly arrived Schaw was startled by 'a number of pigs [that] ran out at a door, and after them, a parcel of monkeys' (1923: 78). This, she writes, 'not a little surprized me, but I found what I took for monkeys were negro children, naked as they were born'. (Sullivan 2012, 302)

Echoing Warren Montag, Sullivan notes how Schaw's experience suggests a "mode of perception, a way of seeing/knowing, that 'enveloped her', that 'saw through her' (Montage 1997: 292)" (Sullivan 2012, 302). The senses, in this particular instance the sense of sight, is co-operational with ontological frameworks that render visible, invisible or otherwise, all objects of apprehension. This runs counter to the still reigning Enlightenment empiricism which continues to insist that our senses help us navigate a static reality in an unfiltered transaction. Even though, with only preliminary reflection, we ***know*** our senses can deceive us, from simple shadows on the wall tricking the eye, to more extreme and racialized instances of injustice, such as the incident in which police officers in the U.S 'swear' they 'saw' a young, black male holding a gun that later turned out to be a phone (Horton and Lowery 2018). Sullivan draws from Roger Lancaster for whom, "as for Merleau-Ponty, perception is 'an embodied social and collective art'" (Lancaster 2003 as quoted in Sullivan 2012, 303) in order to explain how social understandings ***about*** objects work to pre-figure apprehension. In a way, the object is ***always-already*** apprehended. Particularly in the ocularcentric West, it is critical to remember that:

> visuality is the effect and vehicle of sedimented contextual knowledges, rather than a neutral process that provides access to empirical objects/facts, and that 'the realm of the visible, or what is taken as self-evidently visible' (and/or visibly self-evident) is 'the product of a specific form of perceptual practice, rather than the natural result of human sight' (Alcoff 2001: 268). (Sullivan 2012, 303)

Building from this perspective here complements the model of the bodymind I put forth in the Introductory Chapter, that of Elizabeth Grosz's Möbius Strip. Through the Möbius model, I imagine the bodymind to be a site of interrelating organic and non-organic technologies in which consciousness and corporeality are always operating in a complex play of mutual interdependence. However, this 'play' is directly implicated within predominant social schematics. Using this model gives us the opportunity to better understand affective encounters with music listening in the transition between material and immaterial music products. This is also to acknowledge the way in which historically specific listening practices that emerged with material practices helped to cultivate the discursive structures that are in transition today, and why they retain significance. By using this basis, I will assert that the significance of touch, and other sensorial mechanisms such as smell, is still very much a critical aspect of listening in the digital landscape, because of long-entrenched ontological relationships between value and materiality. While Sullivan's work generally focuses on visuality, optics and perceptual schemas constructed through the I/eye relationship to critically examine race and whiteness, I fold that basis into my notion of 'sensorial somatechnics' so that I may focus more on the non-visual aspects of encounter with material objects. In just the same way as our thoughts are contingent upon pre-scripted discursive constructs, so our senses actively reproduce our encounters based on the *limits of knowability* about any particular thing.

Narrativizing the Relationship Between Listening Pleasure and Materiality

The Emergence of Vinyl as a Way of Knowing

Perhaps the first meaningful relationship between contemporary listening culture and objects of consumption begins with the invention of the phonograph. In a narrative sense, it is this relationship that comes to define succeeding expectations between the object and listening pleasure throughout the twentieth century. The phonograph was primarily the brainchild of Thomas Edison, which came into fruition in 1877 (Milner 2009, 34). Phonograph technology works by first capturing the vibration of the original sound through a large horn. That sound, or more accurately, the vibration of that sound, then impacts a diaphragm which is fixed to a stylus. The stylus 'etches' the vibrations analogously onto a suitable surface. In the case of a phonograph, this is a cylinder usually made from a celluloid material. To play this back, the stylus then retraces those vibrations, or 'grooves,' which causes the diaphragm to again vibrate. The resulting vibration is amplified by the horn, playing back what was the original sound (36–37).

I wish to ponder the nature of this specific kind of materiality in relation to sensorial somatechnics, in particular, I relate the practice of 'etching' of vinyl to 'etching' of skin. For example, how we think about how bodies as surfaces that can be 'etched,' I argue, is the first, even though largely unconscious model, that prefigures our sensory connections to the vinyl record. Music, in its nature as an immaterial phenomenon of sound, comes into being by way of the etching process. The etching process manifests the very *thingness* of the musical text; it is this 'coming to matter' that Sullivan remarks on in regard to the "somatechnics of perception" (2012, 302). We might think of this approach to theorizing the vinyl object as a kind of "thinking through the skin," in the vein of Sara Ahmed and Jackie Stacey (2003). Thinking through the skin enacts a kind of politics:

> that takes as its orientation not the body as such, but the fleshy interface between bodies and worlds. 'Thinking through the skin' is a thinking that reflects, not on the body as the lost object of thought, but on inter-embodiment, on the mode of being-with and being-for, where one touches and is touched by others. (Ahmed and Stacey 2003, 1)

In the wake of such thinking, I propose the vinyl record as the inception of the intimate entanglement between tactility and the contemporary musical text. If our perceptual schema maps out preconfigured patterns of knowability, then I suggest the 'etching' of the vinyl skin acts out a mirroring of the potentialities of human skin. I am reminded here of cultural resonances to different versions of skin etching which might mirror the suggestion of surface-etching as a techné through which we come to trace certain objects of encounter. For example, Franz Kafka's etching apparatus from the short story *In the Penal Colony* ([1919] 1948) suggests a kind of ontological mapping in which the skin of the prisoner becomes both materially carved and symbolically territorialized by the colonial machine. In the story, the 'Commandant's invention' for the punishment of a condemned man is an elaborate machine which carves out the sentence of the convict onto the surface of the skin, which over the course of many hours, eventually leads to death. In another sense, tattooing also echoes the relationship between skin-as-object and object-as-skin which produces our relationships to the territorializing process. Tattooing is often cited as a way of 'knowing' one's body, and even a technique through which one can reclaim one's body (Pitts 2003, 56–57). These cultural processes, in which a surface is rendered knowable through the process of cutting into it, presents a certain kind of understanding brought about through our own material embodiment, that is, making the Object in the image of the Self. Borrowing from Merleu-Ponty's notion of intercorporeality, Gail Weiss emphasizes the interconnectedness of "images, imaginary morphologies and materializations" (quoted in Ahmed and Stacey 2003, 5). For Weiss, "the experience of being embodied is never a private affair, but is always already mediated by our continual interactions with other human and non-human bodies" (1999, 5). In a way, to suggest the vinyl as a kind of mirroring process, one that is contingent on the ontological limits and potentials of human skin, we can think about the vinyl listening experience as a mirroring of the relationships we set up between human-on-human encounters. We make objects meaningful through understandings of our own embodiment and objectification; our own materiality helps us make sense of non-human *im*materialities (such as music) in ways that suggests value distinctions.

Further, through the practice of etching, we come to know the vinyl as a process of inscription, which we might say takes on a certain kind of 'aura.' This thinking follows Bartmanski and Woodward's thorough explication of the vinyl object as an "aura-laden object" that is "connected to constellations of other non-human entities that facilitate a series of emotionally charged rituals and experiences on which various communities thrive" (2015, 7). Notions of 'auratic' quality emerge from Walter Benjamin's work, in which aura is defined as "the associations which tend to cluster around the object of a perception" (Benjamin 1968, 186). In application to an understanding of the significance of the vinyl record in its relationship to listening pleasure, Bartmanski and Woodward note that, in Benjamin's words, "To perceive the aura of an object we look at means to invest it with the ability to look at us in return" (Benjamin 1968 as quoted in Bartmanski and Woodward 2015, 21). In this respect, pleasure is entangled with notions and processes of *reciprocity*. In a sense, Bartmanski and Woodward point towards an argument suggesting that the physical object, in its access and physicality, provides a deeper sort of reciprocity than a digital download which cannot be physically handled. However, Bartmanski and Woodward remind us that, for Benjamin, objects such as phonograph records were not considered as auratic because these could be mass produced, and his understanding of 'aura' applied to the singularity of objects. Bartmanski and Woodward set aside Benjamin's qualifier because of the ways in which digital copies of musical texts have reconfigured meaning around the 'singularity' of material copies in a new way:

> Unlike digital downloads (which sometimes have a virtual book accompaniment) the vinyl album encourages an emphasis on the heritage aspects of listening by making its production and playing directly visible and 'to hand' to be read and felt. This makes vinyl as a whole a kind of palpable, durable work of art, i.e. an important companion for ritualised aesthetic practices. (Bartmanski and Woodward 2015, 21)

Therefore, the context of the digitisation has rendered new meaning unto the surface of the vinyl record and reinvigorated our relationship to its materiality.

Cassette Culture and a 'Sense' of Control

The 1960s marked the dawn of cassette technology (Tschmuck 2006, 150) and by the 1970s cassette technology was commonplace in Western contemporary listening culture. Cassettes were similar to the use of vinyl in a material sense, but with one very different and liberating quality—fans could record *onto them*. This suggests an exteriorization of the listening process, a technicity which enables listeners to record from albums, other cassettes and even from the radio. This kind of technicity gave way to entirely new and highly ritualized ways of listening which reshaped what one could do with listening pleasure—that is, capture it in new ways and control it in a way that was not possible before.

One of the technics deployed by the new cassette culture was the production of ritual through game-making behavior as it could be applied to listening pleasure. The putting together of the cassette mixtape (without the ability to cut and paste as we can now through digital programs) had to be 'played out' as a careful game of *order*, based around the rules of song flow, timing, and of course technical equipment. The attraction of game behavior is related to a sense of control because it enacts a world that has rules and order (Huizinga 1955, 10). A sense of control emerges from the success of how one orders and shapes song flow and how the tracks mix and match together—both in the act of creation and the act of listening. The rules and order provided by game worlds are gratifying; they can promote feelings of joy and community, and can be just as (if not more) fulfilling as 'real worlds,' as the predilection for online gaming, fantasy novels and films indicate. Cassette technology, therefore, enabled new ways of creating game worlds in popular music culture and new avenues of escape—a type of shadow reality through song lists. I suggest this produced a new coupling between the listener and the listening experience, in that an added and highly nuanced dimension produces new affective encounters and expectation in the listening field.

The significance of the cassette and the cultivation of listening pleasure through the technicity of cassette technology continues to resonate throughout popular culture, even (or perhaps especially) in the context of digitization. One particular blockbuster film, *Guardians of the Galaxy Volume II* (Gunn 2017), released in 2017 takes the 'cassette' as an aesthetic, which tells us something about the enduring cultural impact of the cassette for the popular music fan. In the film, the main protagonist, Star Lord, saves the Universe from the destructive God known as Ego, just after bolding telling Ego: "you shouldna killed my Mom and squished my Walkman". In fact, the cover art for the film's soundtrack, "Awesome Mix Vol. II," features a cassette and borrows the style of a 1970s color scheme. Similarly, the protagonist in Nick Hornby's cult classic novel *High Fidelity* (1996), later developed for the screen, is obsessed with perfecting the playlist. Again, in the film *Cuban Fury* (2014) the male protagonist slaves over the perfect song order for a mixtape intended for a woman he aims to woo. While MP3s have greater convenience than cassettes, the cassette has been refunctionalized as a product with *emotional* value through which one might couple one's listening experience.

This emotional narrative is explored by the short film *You Need to Hear This*, in which three cassette tape aficionados "explore what drives their enduring love for the cassette tape" (Kenny 2013). During the film, Jen Long, founder of cassette-only label Kissability, explains that when she was younger she "had this cassette recorder thing, it had a microphone on it so you could record yourself, and it came with this yellow tape … one side was blank so you could record over it…" (Kenny 2013). In this instance, materiality confers a connection with the music in a deeply personal way—which transmutes into, and is fluid with, subcultural listening practices. In comparing digital playlist formats versus those on cassette tape, Henry Rollins decries that, "digital is almost disingenuous" (Taylor and Petzold 2012). It is also worth noting that according to the Official Charts Company, in the UK, cassette sales grew in one year (from 2017 to 2018) by 90 per cent (Copsey 2018). What this constellation of cultural instances demonstrates is the reinscription of symbolic currency onto the cassette as an object of culture in the wake of digital downloads and streaming services. This trend indicates a reinvigoration of materiality in popular music listening culture.

Further, as a result of the easy technological processes and inexpensive equipment required for mixtape production, I would also suggest that the cassette, as an object, also helped to mobilize social listening practices. In the U.S., in the 1980s and particularly in the underground hip hop scene, DIY cassettes were integral to defining lines of empowerment in a growing industry ever-increasingly modelled on technocratic neoliberalism. Anthony Kwame Harrison writes that the:

> distinct DIY underground hip hop movement that emerged in California (around local scenes in Los Angeles and the Bay Area) [was] a response to the commercial rap music industry's unwavering commitment to gangsta rap imagery and themes. (Harrison 2006, 285)

In the counter-capitalist ethos of mixtape DIY, music scenes mark out affective lines which demarcate boundaries of independence from corporate interests. The cassette mixtape, in this sense, becomes a site unto which feelings of empowerment and territorial strategies meet for both listener and creator. For Harrison, it was the cassette tape that supported "hip hop's peculiar standing at the crossroads of anti-corporate subcultural movement and cottage cultural industry" (2006, 288) in the Bay Area underground scene. Harrison argues that the DIY nature of the cassette enabled a defense against the mainstream because analogue taping could function in ways that digital formats, like the compact disc, could not. For example, one way in which individuals expressed their feelings about artistic integrity in this scene was via the "crudeness" of the cassette tape—as opposed to a more professionally produced technology such as a compact disc. In this sense, "Crudeness can be thought of as an alternative aesthetic orientation that makes use of imperfections and sonic disjunctures to convey a sense of underground hip hop authenticity" (289). In Jennifer Fisher's line of thought, we might describe this as the way in which a "tactile aesthetics strives to 'deepen the sense of subjective embodiment' by foregrounding the function of bodily modes of experience" (Fisher 2006 as cited in Hansen 2006, 110). As a result of the insistence of the physical aura, Harrison views "audiocassette tapes as unique technologies that simultaneously embrace the progressive politics of subcultural inclusion while defending subcultural boundaries against mainstream co-optation" (283). In this instance, actors in local scenes form an emotional attachment to analogue technologies which are used to express the body-politic especially along racialized lines of embodiment. Here, again, we note the ways in which the physical nature of musical commodities are not only neutral containers for the exchange of musical texts, but in fact operate as a techné which extends and recapitulates the limits of encounter between the fan and the corporate system, cultivating a listening culture through lines of connections with material objects.

The Compact Disc, Mobile Music and Proprioception

The compact disc ushered in the age of digital material formats with the introduction of the WAV file. The compact disc was released across Europe and Japan in 1982, and across the U.S. in 1983 (Pohlmann 1989, 12). The CD was marketed as a 'superior' format which could not be surpassed. In fact, the industry was so intent on establishing the CD as the premier listening format that, in some instances, vinyl LPs were physically taken off shelves (Coleman 2009, 141). However, even at the peak of the compact disc age in the late 1980s, MP3 technology was already being developed by several different groups (see Sterne 2012). By the mid-1990s an MP3 file standard was settled in the world of global communications.[2] MP3 file sizes are so small (in terms of data) because the technology takes the original music file, with all its reams of information, compares it to a mathematical representation of the gaps in our hearing (Sterne 2012, 1–2) and cuts out what we cannot 'really hear.' By cutting out the sections of the audio that we cannot really hear the file can be reduced to around twelve per cent of its original size (2). As a result, music files can be rapidly uploaded, downloaded, stored and shared online. It has made digital programs like iTunes possible and completely revolutionized mobile listening, and with it, the limits of listening pleasure in relation to psychogeographical encounters.

To move this discussion into the field of sensorial somatechnics, I focus on the sensory system of proprioception in relation to the uptake of iPod use in the early 2000s. Proprioception refers to "the sense of position and movement of our limbs, the senses of muscle force and effort, and the sense of balance" which "allow us to carry out our tasks successfully, without thinking" (Gandevia and Proske 2016, para. 2). To situate proprioception in a 'sensorial somatechnics,' I imagine these faculties as the technicity which defines our very place in the world and which helps us to connect our body to other experiences, such as music listening. In the words of Charles T. Wolf, proprioception gives us a way of understanding the "priority of dynamic embodied activity over isolated 'mental' and 'physical' regions" (Wolf n.d.). For Wolf, rather than asking "How can a brain accomplish reasoning?", the question becomes "How can a brain have experiences?", that is, "What is it like for a brain to be embodied?" For Michael Bull this kind of embodiment is of particular importance in the exploration of iPod use, especially as a technology of interacting (or shielding oneself) from the urban landscape. For example, in Bull's empirical research, he examines the commute of an iPod user, noting that:

> In the often-repressive 'realm of the eversame' (Adorno, 1976) or the 'ever-always-the-same' (Benjamin, 1973), the iPod user struggles to achieve a level of autonomy over time and place through the creation of a privatised auditory bubble. […] In this de-routinisation of time lies both the unalloyed pleasure of listening but also the management or control of the user's thoughts, feelings and observations as they manage both space and time. (2005, 344)

[2] MP3 is short for MPEG-1, Layer-3. MPEG refers to the Motion Picture Experts Group, a consortium of engineers (Sterne 2012, 829).

The very control of one's environment is perched on one's ability to move through said environment. The depth of such experience is cultivated through the complementary functioning and even 'skill' of iPod use, as it moderated with proprioceptive boundaries. Susan Broadhurst and Sara Price discuss the critical links between these sensual encounters and the use of technologies, worth noting in full:

> There is also the phenomenon of what the direct use of physical instrumentation does to our innate sense of being embodied (sometimes called proprioception). When we build such instruments we consequently project around ourselves a mediated world, in effect our 'somatic limits' spread to include our created tools. This phenomenon was noted by Heidegger and Merleau-Ponty. Rather, than being separate from the body, technology becomes part of that body and alters and recreates our experience in the world. In many contemporary digital artworks, the human body is shown in flux, a body where contacts are made not physically but electronically where cutting edge multimedia effects explore the inherent tensions between the physical and virtual. In social science contexts contemporary digital technology changes communicational resources and communicational space, bringing new forms of interaction and the potential for new creative spaces, and new forms of 'making meaning.' (2017, 3)

Of course, the Walkman and personal stereos have existed since the late 1970s, however, these technologies limited music choice, content order, and the manipulation of settings. Conversely, digital music players store thousands of songs which can be arranged and listened to based on a variety of configurations (Bull 2005, 343–344). As a result, "Technologies like the Apple iPod produce for their users an intoxicating mixture of music, proximity and privacy whilst on the move" (343–344). The digital and mobile playlist technology gives users not only an "intoxicating mixture" of music and emotion, but often it produces a sense of control over the very nature of time and space and their introjection both physical and mentally into that time and space.

The Discourse of Authenticity: Materiality as a Discursive Construct

As I have expounded in the introduction to this chapter, our sensorial somatechnics are constituted by an active production of meaning-making mechanisms and that filter, channel, figure, and disfigure our transactions with the phenomenal world. Here, I explore the ways in which the construct of materiality as a form of 'authenticity' shapes and manages listening culture and points us towards certain kinds of listening behaviors. In analyzing music listening discourse, there is a sense that material listening practices can provide a more 'authentic' experience than virtual listening practices in the context of digitization.[3] Twentieth century constructions of authenticity and materiality can be viewed as a legacy from Renaissance ideals, in which the original artwork (usually a painting or sculpture) was consecrated as the 'authentic' version, even while exact replicas could be made (Butler 2006, 469). From this basis, Walter Benjamin theorized that, "The presence of the original is the prerequisite to the concept of authenticity" (1968). Benjamin explained that, in the age of mechanical reproduction, "Even the most perfect reproduction of a work of art is lacking in one element: its presence in time and space, its unique existence at the place where it happens to be" (1968, 214–216). For example, the Mona Lisa is kept behind glass in the Louvre while the exact image of the Mona Lisa is printed on tea towels in tourist shops outside. The image is the same but the value is dependent on the painting's materiality in 'time and space.'

The value of authenticity in popular music discourses works in similar ways and the significance of this cannot be overstated here. Hugh Barker and Yuval Taylor characterize popular music culture's preoccupation with authenticity as a 'quest,' a search for that which is truly genuine (2007, viii). However, this quest is built around certain rules. For example, even though a vinyl album is still a reproduced and reproducible object, the material fact of the vinyl lends 'weight' (both literal and figurative) to the ideal of authenticity, as opposed to the *virtual* reproducibility of an MP3. I suggest here that this construct retains significance as a direct consequence of the material relations built from these twentieth century (and even older) practices, in which authenticity was bound to touch and materiality.

As a result of the relationship between authenticity and materiality, the material product, therefore, should be approached both discursively and 'somatechnically'. My argument here emphasizes that discourse does not operate outside the scope of sensorial somatechnics. What I mean is that the significance and affective value of the material product is not only

[3] Elsewhere, I have explored this conversation by way of a cyberethnography of Reddit users (Glitsos 2017).

shaped by institutional language—from the discourses of the creative industries to the ideology of neo-capitalism—but also by the forces of material relations as they relate to the phenomena of touch and tactility that is experienced by the consumer and become narratives of encounter. For instance, music blogger Simon Sweetman says that "there's a physical connection with vinyl. You are forced to interact with it; it's tactile" (2011). In a sense, the corporeal exchange shapes the discursive site of the material product and vice versa, in the Möbius inflection which generates concomitant and enfolded experiences of both mental and bodily intellects.

The significance of touch as a product of twentieth century material relations can be exemplified by the new approaches to psychosomatic phenomena that emerged during the early to mid-1900s. For example, Sigmund Freud's deployment of touch as a "technical tool," or what we might read as a somatechnique, in the process of patient therapy (Bartole 2011, 379) sought to concretize the intimate link between the psyche and sensorial phenomena. When Freud would encounter resistance from a patient he would exert gentle pressure with his hand on the patient's forehead, while explaining to the patient that the effect of the touch would conjure thoughts and memories. Using the power of suggestion and the sense of touch, the patient was then directed to articulate their thoughts (379). As Tomi Bartole explains, the purpose of this action was to redirect the patient's awareness, therefore leaving a space through which could enter unconscious or buried thoughts (380). Freud's technique deploys the body's somatechnics as a mechanism in which touch relays and builds cognitive and affective materials. The exchange in material music listening practices with textural objects also relays cognitive and affective materials—feeling the weight of an LP in one's hands, flicking through CD liner notes, and so forth—have become bound to unconscious pleasure or emotional release. Consumers *expect* this listening experience to produce affective exchange that differs from virtual experiences, and this expectation is a powerful force in the investment of listening to music on material formats.

Further, materiality can be read in relation to the power of cathexis. Cathexis can be understood in terms of somatechnics because the process suggests that psychic energy is translated into an emotional connection with an object. This process works as a kind of technicity that can move energies to and from the body's inner delta. For Freud, the *valeur affective* of an object draws its power from the individual's ever-present longing for the maternal object (Laplanche and Pontalis 1973, 65). This "cathexis of longing," or *Sehnsuschtbesetzung*, suggests a deep never-to-be-recovered desire for that missing object (65). Similarly, in an 'age of obsolescence,' as the digital context is sometimes called (Fitzpatrick 2006), a sense of longing for traditional music listening practices, and even for the objects themselves, is often cited in popular music discourse.[4] Russell Belk approaches the relationship between emotionality and materiality in a similar sense. He writes that "digital possessions lack the soft tactile characteristics of clothing and furniture that make it possible to almost literally embed our essence in such possessions" (2013, 480). While Belk's approach may suggest an 'inherent-ness' in emotional responses to materiality, I think it is more pertinent to focus on the ways in which historically specific modes of music listening have been deeply embedded in popular music listening culture, both discursively and through sensorial somatechnics. The desire, and the longing, for music *became* rooted in the physical nature of the listening experience across a variety of popular music practices as they were bound to twentieth century understandings of the authenticity of materiality.

'Libidinal Energization of the Commodity'

Material products can also act as a kind of libidinal prop, and in some cases are even fetishized in popular music culture. A prop is that which stands in for something else, and fetishism implies the practice of worship, originally referring to the worship of abstracted gods by 'primitive' peoples (Kaplan 2006, 17). In contemporary terms, fetishism suggests an erotic aspect and generally refers to the "displacement of erotic interest onto an object, such as a shoe" (16). For Louise Kaplan, fetishism is approached as a "strategy," which is the actual *process* whereby one fetishizes an object. For Kaplan, this process emanates from "the need to transform something unfamiliar and intangible into something familiar and tangible" (16). For my purposes, I read the fetishization of the material music product as a process by which one makes attainable the unattainable, that is, how one might materially manifest something abstract such as music. Music itself is a quasi-

[4] For example, pop culture website Buzzfeed laments late twentieth century music listening in the article "35 Music Experiences You'll Never Have Again" (2013), which includes activities like "waiting outside record stores for midnight releases". See Tanner Greenring and Jack Shepherd. 2013. 35 Music Experiences You'll Never Have Again. *Buzzfeed*, August 16.

material phenomenon, because a song cannot be held, touched or given substance *but* through the materiality of the music product. Through material relations music can take on substance and is endowed with another dimension in time and space. That is, the material object is the prop for the immaterial musical text.

In the West, twentieth century consumption practices cultivated a fetishization of material products (Lunning 2013, 7) that continue to resonate in the context of contemporary digital culture. Twentieth century commodity fetishism arose from an infinitely complex set of historical and social circumstances, too long to detail here.[5] However, it is important to note that this history extends at least as far back as the development of the European mass market during the second half of the nineteenth century (Stratton 2001, 26). In Stratton's text, *The Desirable Body*, he cites Stuart Ewan who explains that, by the introduction of assembly-line mass production "'excessiveness replaced thrift as a social value. It became imperative to invest the labourer with a financial power and a psychic desire to consume'" (Ewan 1976 as cited in Stratton 2001, 32). Institutional discourses—of advertising and industry—as well as state apparatuses governed by free-flow economics, would enable the unmitigated growth of consumer culture. As Stratton goes on to explain, "commodities became, themselves, constructed as fetishes" which was a process that "involved the libidinal energization of the commodity" (32). This is the process whereby material listening objects, such as the vinyl or cassette, became imbued with a sensorial somatechnics. For example, faculties of sight, touch and smell are coupled by the energizing forces of commodity economics through which notions of pleasure and ownership of music are tied to the physical nature of the object.

Popular music and commodity fetishism intersected from the post-war period onwards and manifested in such moments as the 'hippie' trend of the 1960s, the "fetish fashion" of 1970s punk, and the mainstreaming of punk fashion in the 1990s (Lunning 2013, 7). Popular culture, fashion, sex, gender and music became inextricably during the post-war period and many commodities were fetishised in the process. Aggressive marketing and the rise of hyper-consumerism gave way to a lifestyle culture. As Kaplan writes:

[5] Jon Stratton covers this history in detail. See *The Desirable Body: Cultural Fetishism and the Erotics of Consumption* (2001).

> Material objects that are regarded with extravagant reverence and sought after with a compelling, 'I've got to have it,' are fetishes. Such items could be almost anything—chiffon scarves or Manolo Blahnik stiletto sandals, Prada handbags or Chanel jackets… Or, to leave the world of fashion, there are kitchen utensils, or cell phone attachments, iPods. (2006, 17)

In popular music culture, music fans worship rare vinyl or limited editions, or even their favorite CD, perhaps put on display by framing and hanging on a wall. The listening experience is permeated by the intense connection to its correlate object. The intangibility of listening pleasure is transmuted into a concretized form that can be touched, stroked, held and even shown to others in three-dimensional form.

Connected to the strategy of fetishization is the way that music fans can exhibit extreme behavior in hunting down and buying rare copies, limited editions or further paratextual materials such as artist merchandise. Special editions are still popular, and many are still released with textured covers or embossed sleeves. In fact, as one extreme example, the Wu Tang Clan's *Once Upon in Shaolin* (2015) has only one materially-existing copy (see Green 2017). The material product is therefore constructed as an experience that provides *supplementary* dimensions to a product that previously could be experienced in only one dimension. The listening experience, in this respect, is also constructed as an emotional-sensual enterprise which manifests a host of sensorial phenomena driven by the desire to connect in a deeper, perhaps more personalized, way. Consumption modalities shape these practices, as Shuker explains, "Recording industry packaging practices have created a number of collectibles, including picture discs, picture sleeves, boxed sets and 'promos'" (2010, 57). Consumers use these artefacts as strategies to individualize the music experience. James Clifford writes, "At a more intimate level, rather than grasping objects only as cultural signs and artistic icons, we can return to them … not as specimens of a deviant or exotic 'fetishism' but our own fetishes" (1988, 229). The material product functions as a critical object in the context of digital music because it provides an aspect of the listening experience—namely, its material sensuality which can be fetishized on some level and in idiosyncratic ways—that cannot be as readily reproduced through virtual modes.

It should be noted, however, that the process of fetishization is not to suggest that the material object provides an experience that is *inherent* to music, or an experience that is objectively more authentic. As Kaplan points out, fetishism is associated with falsity (2006, 17, 19). The word 'fetish' comes from the Portuguese, *feitico*, meaning false (17). One worships the *object in place* of the abstraction, that is, a feeling or sensation (originally referring to the worship of abstracted gods by 'primitive' tribes). The abstract thing, whether it be a sexual fantasy or an emotional release, cannot be easily accessed nor easily articulated by the individual—but the object *can* be accessed at will: "A fetish can be held, seen, smelled, even heard if it is shaken, and most importantly it can be manipulated at the will of the fetishist" (20). The object stands in for, or provides greater substance to, the immaterial experience. Similarly, in looking at the ways in which listeners describe their experience of traditional listening modes, the materiality of the products stands in for the emotional release or pleasure of the listening experience; they become bound together in psycho-sensual patterns of practice.

In the framework of sensorial somatechnics, these psycho-sensual patterns of practice present a key aspect in understanding the place of touch and material relations in the age of immaterial music. In the scope of music listening, sensorial faculties translate the physicality of the experience in more complete terms than virtual experiences alone. I return to the work of Elizabeth Grosz to highlight the functionality of touch in the play of both sound and fetish. In Grosz's articulation of the Möbius body, "the skin and the various sensations which are located at the surface of the body are the most primitive, essential, and constitutive of all sources of sensory stimulation" (1994, 35). The pleasure of touching is translated through material relations and can support the listening experience in a way that seems to alter and attenuate those barriers between inner and outer dimensions. As Grosz continues:

> The information provided by the surface of the skin is both endogenous and exogenous, active and passive, receptive and expressive, the only sense able to provide 'double sensation' … [which means] the subject utilises one part of the body to touch another, thus exhibiting the interchangeability of active and passive sensations, of those positions of subject and object, mind and body. (35–36)

The flux and flow of intensities decorate the listening experience through the interaction with materiality. The listening experience is modified and shaped by the listener's concrete presence in three dimensions. This is not to suggest that material listening has superiority over immaterial formats. Both material and immaterial listening practices provide different forms of pleasure and/or conveniences that cannot be quantified as more or less valuable than the other. Rather, this is to suggest that the material product provides aspects that the virtual cannot and therefore the material product still retains a critical, albeit redefined, function in listening culture. Material products do not function as the only access to music listening now. Instead, material products are a singular listening style deployed in order to produce different emotional and sensory experiences that privilege the critical function of touch.

Sensual play facilitates a variety of strategies in the fetishization of music: fans can hold the creative *lifeforce* of an artist in one's hands and listening then becomes also touching, smelling, feeling as well as listening. For example, fans of heavy metal can purchase CD covers signed in real blood: Australian metal band Malignant Monster sign their CD sleeves in the blood of the lead singer and seal the sleeve with a wax insignia. This suggests (a fairly intense) desire to connect with the artist producing the music to which one is listening in more dimensions than aural alone, and it also demonstrates the way materiality is implicated in that desire. In the case of Malignant Monster, the use of blood on the CD sleeve resonates with the act of listening, particularly to metal music which is often constructed as an 'outsider' genre, in that both suggest a degree of transgression. In the same way that sound permeates and moves through bodily borders, so does blood, albeit in a different way. As Grosz explains, "Body fluid attests to the permeability of the body, its necessary dependence on an outside, its liability to collapse into this outside (this is what death implies), to the perilous divisions between the body's inside and its outside" (1994, 193). Blood both illustrates the closeness of the inner and outer worlds, while at the same time drawing attention to the wide chasm between them.

Further, in this example, the blood symbolizes and solidifies a bond between listener and artist, which can be achieved through the material exchange. For example, the use of a fluid like blood has Biblical connotations, in which Jesus Christ asked his disciples to 'drink of his blood' in order to seal the covenant—the ultimate bond. In a similar sense, blood is used symbolically, as well as physically, to point to lines of heritage and familial ties. In quite another sense, blood takes on sexual connotations, such as in its use in vampire narratives to suggest a visceral link between two individuals (Seed 1985). In all these scenarios, blood is deployed as a substance that can confer a powerful connection. The case of the Malignant Monster CD sleeve exemplifies the way material culture can play with such themes in physical formats. This sentiment is echoed in popular discourses in which music listeners have claimed that physically having an album can help facilitate a connection to an artist or listening encounter. Different modes of listening connect consumers to the artist in the act of the listening experience. This is not to contend that all, or even most, contemporary listeners react this way to the immaterial context. Rather, this is to illustrate how some affective transitions take place through a sensorial somatechnics, that is, a technicity which renders the listening experience as knowable in a variety of distinct and overlapping ways. The material product can often (but not always) produce a more visceral experience than the MP3 format and therefore brings forth the sensual play of touching (the Other) into the act of listening.

The role of materiality is also significant in the cultivation of connection to other listeners. It should be noted that the digital context has produced new ways of connecting to other listeners in the online community. However, I focus here on the role of the material product in producing subcultural associations. For example, listening to vinyl is often thought about as a niche, community-organized practice. In one of the most comprehensive texts on vinyl record collecting, Shuker writes that, "The 'social practices' [...] shared by other record collectors, presents an interwoven narrative of desire and identification, alongside notions of cultural and economic value, which characterize many collectors' accounts of their passion" (2010, 10). For example, music blogger Simon Sweetman writes about the social process he uses to generate 'first listen' experiences, which he calls "the crate game" (2011). In the "crate game," Sweetman and his partner store a crate of brand new, unplayed and hermetically-sealed albums at home, and when someone comes to visit they are given the honor of selecting an album from this crate in order to 'first-listen' together. Sweetman himself describes this as:

> [A] ritual for me and Katy; we both enjoy seeing what gets chosen—sometimes that's as interesting, in a way, as the actual album. The process of elimination—from what gets shortlisted—can be quite intense and it's often amusing. In the end, after all the careful planning, it might come down to a whim, a prettier cover, a sicker/weirder image, a name a person has never heard or a record that reminds them of their childhood. (2011)

Sweetman and Katy translate the processes of subcultural membership—knowledge about music—into the pleasure of music listening through exteriorization of the musical text into material object, the basis of techné.

On this, David Howes notes that "the meanings and uses people discover in or ascribe to [material artefacts] in accordance with the sensory order of their culture or subculture" (Howes 2006, 166). In this case, the vinyl artefact lends itself, as an object in time and space, to the practice of the subcultural ludic ritual based on material artefacts.

Memory and Ownership

Touch can also imply a sense of ownership and has become another way in which consumers redefine their experience of owning music in the context of digitization. Many consumers may play, stream, and exchange MP3s online, but will also suggest that this does not necessarily constitute owning a music collection through which emotional connection is constructed (Glitsos 2017, 68–69). Music blogger Luc Duval shares his experience with materiality:

> I imagine it would be quite difficult to play records without feeling a more palpable and personal awareness of, and connection to, the music that one listens to. I even feel a stronger sense of ownership of the music that I have on records than the music I have stored on my computer. (2011)

Duval is articulating a very specific experience in music listening; a sense of ownership over the listening experience itself. This is not to imply that all music consumers feel that materiality confers ownership, however, it is to suggest that for some consumers the material product and the *notion* of owning are related and therefore produce positive listening experiences when one deploys those products. Ownership has long been associated with having something tangible in one's possession (as per the old adage that "possession is nine points of the law" [U.S. Legal 2001]) and here individuals similarly point to a sense that a physical copy has a different quality than having access to a digital collection. Whether or not we can characterize these experiences as 'real' or 'justifiable' is not the point of course. What is crucial is that the individual experiences her or his listening pleasure *as* different based on the constructs at play in relation to ownership. Feelings are subjective in their very nature and therefore there need not be an objective correlation between the fact of materiality and sensorial somatechnics.

A similar sentiment is related to the new meanings around collecting material music products in the context of virtual listening practices. Collecting material music products, particularly vinyl records, emerged as a significant and ritualized practice in popular music culture (Shuker, 2010, 2012). In *Wax Trash and Vinyl Treasures* (2010), Shuker outlines the contexts which enabled the practice of vinyl collecting to flourish:

> During the mid-to-late nineteenth century, a mix of capitalism and consumerism, increased leisure time, disposable income and nostalgia made collecting a significant aspect of the social identity for the new middle classes of Europe, Britain and its colonies, and the United States. Record collecting as a social practice was a logical extension of such activities. (2010, 3)

In a complementary discussion, David Beer explores several concepts behind the record collecting phenomenon and suggests the practice performs several functions that immaterial music cannot replicate in the same ways. For example, Beer explains that he has compact discs in a collection that still have "stickers on the case from a small independent record store in Derby city centre" (2008, 75). In Beer's words:

> Collecting is the accumulation of a form of material biography that reveals things about us, about our life trajectories and histories, and about the social and cultural movements, moments and events that we have lived through or that we find connection with [...] the physicality of the collection is an integral part of our relations with it and the identity constructions it facilitates; this is both how we think of ourselves and how we wish to present ourselves to those that visit the private places where we exhibit our collections. (2008, 75–76)

The most critical aspect in this excerpt from Beer, for my purposes, is his insistence on the *physicality* of the collection in terms of cultivating pleasure. Collecting material music products is about collecting experiences and ways of feeling about those experiences in their function as identity-making instruments. As an extension of this process, listening to those products is a way of accessing those experiences and markers of identity, which are evidently tied to these material artefacts.

One of the unique aspects of vinyl records is that the vinyl and the paper cover of the artwork is porous and, as a result, retains smell. The sense of smell is an incredibly powerful somatechnique (see Classen et al. 1994) which relies not only on human faculty but on the physical nature of the object at hand. Thus, there emerges again a reciprocal and sensual relationship between the listening object and the 'feeling subject,' through which listening pleasure is underwritten. According to Constance Classen,

David Howes, and Anthony Synnott, "odours affect us on a physical, psychological and social level" (1). In their research, Classen and Howes recount the story of a man who lost his sense of smell and described this event "like being struck blind" (1). The man in question noted that: "Life lost a good deal of its savor ... You smell people, you smell books, you smell the city ... My whole world was radically poorer" (1). In a similar sense, vinyl records and their covers are ultimately connected with smell which *enriches* the musical encounter in many ways. Music commentator Jason Lief defines his relationship to vinyl through smell:

> I would always open up the package, pull out the lyrics printed in tiny letters, and then I would take a whiff. Different labels (so it seemed) had different smells. So Stryper smelled different than Van Halen and Whitesnake had a different scent than Dokken. It was a material experience through and through, from the grooves to the jackets and liners—the smells and bells of rock n' roll. (Lief 2014)

One can also find an entire thread on Reddit dedicated to the experience of smelling vinyl records (WeAreEvolving 2018). These kinds of discussions point to the implication of sensorial somatechnics in cultivating the nature of the listening experience as it is made possible through the material object.

The connection between the senses and memory cannot be overstated either. Music fans can deploy material products in order to construct and reconstruct memory, and produce pleasure related to acts of remembering, particularly in ways that consumers cannot do with immaterial formats. Grant McCracken explains that "goods help the individual contemplate the possession of an emotional condition, a social circumstance, even an entire style of life, by somehow concretizing things in themselves" (1988, 110). This idea also resonates with Pierre Nora's work on *les lieux de mémoire* (1989). Nora's work was written in the late 1980s, however, it still retains relevance here because the 1980s was also a time of great transition in media, with the introduction of the home computer, compact discs, the personal Walkman, and fax machines. Nora discusses a cultural anxiety around the "collapse of memory" in the postmodern age, which in turn catalyzes "consciousness of a break with the past" (7). He writes: "There are *lieux de mémoire*, sites of memory, because there are no longer *milieux de mémoire*, real environments of memory" (7). Beer, too, explains the way that connection to musical products is not just a case of connecting with a physical item, rather, it is the connection to the narratives that emerge "through the unique relations that the collector has with that particular object" (2008, 75). Beer describes the emotional process of looking through his compact disc collection, complete with the stickers and labels from each store:

> These labels bring to mind memories of times spent, during my youth, rifling through the shelves of the store, an experience that is recalled merely through the presence of these sticky labels. These labels also have the price and short messages from the record store staff (about the content of the CD) written on them; this recalls the relations and moments of connection between me and the staff over shared interests in the music. (2008, 75)

Memory and sensorial phenomena are profoundly linked. For Shuker, material culture informs a "strong connection" between music products and memory (2010, 53). Material products have been constructed to interact with memory and emotion in different ways than immaterial practices have been constructed because they 'take up' physical space and inhabit physical form (in much the same way as the human body does). Material products can inform the listening experience as idiosyncratic and personal by using the structures of material culture as a site of meaning and meaning-making.

Material music products also enhance the processes of memory in ways that other material products cannot because of the relationship between music and time. Catherine Strong explains that, "Music provides a different type of access to memory than other artefacts of the past, such as photographs, because of the way it moves through time itself as it plays" (2015, 421). Strong draws from Tia De Nora's work to emphasize the unique capacity of music to structure processes of remembering. For instance, a photograph or postcard is static, these are products that exist in time and space, but do not appear to move through time *with* the user. A song, however, forms a direct connection to time as it moves through it. In De Nora's words, music "provides a device for unfolding, for replaying" (De Nora 2003 as cited in Strong 2015, 421), referencing the way that music appears to the individual as a companion through time and, in a sense, travels with them through the shifts between pasts and presents. A song appears to trace the untraceable; the slippages of time can be marked through a melody or well-worn beat. Of course, the relationship between music and time is not tied to its materiality, however, coupled with the ideas above in which I illustrate the connection between material culture and memory, the material music product works as a device for processes of remembering, which heightens and individuates listening in nuanced and distinct ways.

However, though material listening practices may be distinct from MP3 listening practices in terms of connecting with memory, that is not to say that material products are superior. In fact, many individuals can and do deploy a range of music mediation devices in order to take pleasure in the relationship between music and memory, such as the exchange of YouTube song clips between friends on Facebook that serve to maintain personal connections to shared experiences, as just one example. As Andy Bennett explains, one of the "palpable effects" of "mediated memory has been a *diversification* of the ways in which the past is remembered and represented" (Bennett 2010, 246; original emphasis). Bennett points to the complexity of how memory and music might come together in the contemporary digital framework, where one can more easily look up a song on YouTube than dig through an entire record collection sitting in another room where the material format of the song is stored. Mediated representations of memory through music do not supersede the material processes of remembering, but can complement it, and in some ways even augment it. As Bennett continues:

> Rather than struggling with a narrowly codified, not to say monolithic, set of representations, late modern individuals may find within the realms of mediated memory multiple frames of reference through which to organise and rehearse their own personal memories of the past. (2010, 246)

As music becomes reframed by digital contingences, so too does the individual's investment of memory, in both private and shared spaces, and material and immaterial formats. The exponential expansion of modes of listening are then 'mixed and matched' together in playful ways.

In his article, Beer goes on to question the residual effects of the shift from owning to the more common practice of 'streaming' or 'sharing'. He asks, "if the explicit physical identity forming and material biography properties of the music collection are removed (or at least realigned) from the domestic space and reinscribed upon a virtualized and mobile digital file, what are the consequences?" (2008, 77). For Beer, marketing discourses intersect with new listening practices in order to reshape what collecting itself means in order to privilege the "iconic interface [of the MP3 player] and the veneer of simplicity" (71, 74). I would suggest that, in addition to Beer's account, the significance of collecting has also been somewhat superseded by the significance of *sharing*. Volker Grassmuck calls this the "sharing turn" (Grassmuck as cited in Belk 2013, 1596) where the traditional notion of "you are what you own" is recapitulated as "you are what you share" (1599). I suggest that the new sharing culture is manifest in the many peer-to-peer download websites, streaming services and online platforms available to music fans. What has become critical is the collecting of 'sharing experiences,' particularly through the countless platforms to which a listener can register. Registering for a site, such as Spotify, where one can expose their playlists and critique the tastes of others, is almost a ritual of collecting itself. Users can store each login detail for each different sharing experience, and rotate visits to each sharing site, in a kind of ritualistic play that extends sharing pleasure. Far from constructing the process of digitization as an empty and 'soulless' expedition, instead, there is a sense of discovery and therefore hope and excitement. Music fans can find new songs and new avenues of music listening. This is not to say that virtual possessions can easily take the place of material possessions. Instead, as Janice Denegri-Knott and Mike Molesworth (2010, 110) suggest: "Virtual goods occupy a liminal category between the material world and the imaginary world" (479). Digital experiences have not replaced traditional ones, rather, new listening experiences create an entirely new space for the grounds of engagement and what it means to feel emotionally connected to listening experience. Rather than connecting with tangible items, individuals connect with the activity of sharing and as they listen and stream online music, individuals invest psychic energies into that experience.

Conclusion

The bodymind emerges in and from the world, bridged by sensorial somatechnics which serves to connect the subject to listening pleasure. In this chapter, I have thought through some particular sensorial techniques: olfactory, tactile, proprioceptive and visual. By doing so, I connect the world of the senses, or what David Howes might call "the empire of the senses" (2005), to the pleasure of listening as it produced through technicity.

I began this chapter with an assertion borrowed from Dominik Bartmanski and Ian Woodward's research on the recent upsurge of the vinyl record in which they note that the vinyl takes on a renewed significance in an era of immaterial music listening practices, such as streaming services and download culture. Material music culture, like any practice, is discursive, and as I have noted, historically specific listening practices helped to cultivate the constructed relationships between the pleasure of listening and the authenticity of the material object. These constructs led to the concretization of materiality and listening pleasure, and thus defined the emergence of vinyl as a way of knowing that pleasure. Moreover, I have further built on Bartmanski Woodward's premise by engaging a 'sensorial somatechnics' in order to note the singularity of material practices and how they have been reinvested with particular import in digital contexts.

References

Ahmed, Sara, and Jackie Stacey. 2003. *Thinking Through the Skin*. Hoboken, NJ: Taylor & Francis.

Barker, Hugh, and Yuval Taylor. 2007. *Faking It: The Quest for Authenticity in Popular Music*. 1st ed. New York: W. W. Norton.

Bartmanski, Dominik, and Ian Woodward. 2015. The Vinyl: The Analogue Medium in the Age of Digital Reproduction. *Journal of Consumer Culture* 15 (1): 3–27.

Bartole, Tomi. 2011. Freud on Touch: Thinking Sexuality in Anthropology. *Esercizi Filosofici* 6: 376–387.

Beer, David. 2008. The Iconic Interface and the Veneer of Simplicity: MP3 Players and the Reconfiguration of Music Collecting and Reproduction Practices in the Digital Age. *Information, Communication & Society* 11 (1): 71–88.

Belk, Russell. 2013. Extended Self in a Digital World. *Journal of Consumer Research* 40 (3): 477–500.

Benjamin, Walter. 1968. *Illuminations*. New York: Schocken.

Bennett, Andy. 2010. Popular Music, Cultural Memory, and Everyday Aesthetics. In *Philosophical and Cultural Theories of Music*, ed. Eduardo De La Fuente and Peter Murphy, 243–262. Leiden: Koninklijke Brill.

Broadhurst, Susan, and Sara Price. 2017. *Digital Bodies: Creativity and Technology in the Arts and Humanities*. London: Palgrave Macmillan.

Bull, Michael. 2005. No Dead Air! The iPod and the Culture of Mobile Listening. *Leisure Studies* 24 (4): 343–355.

Butler, Beverly. 2006. Heritage and the Present Past. In *Handbook of Material Culture*, ed. Christopher Tilley, Webb Keane, Susanne Küchler, Mike Rowlands, and Patricia Spyer, 463–479. London: Sage.
Clan, Wu Tang. 2015. *Once Upon a Time in Shaolin*. Cilvaringz/RZA.
Classen, Constance, David Howes, and Anthony Synnott. 1994. *Aroma: The Cultural History of Smell*. London: Routledge.
Clifford, James. 1988. *The Predicament of Culture: Twentieth-Century Ethnography, Literature, and Art*. Cambridge, MA: Harvard University Press.
Coleman, Mark. 2009. *Playback: From the Victrola to MP3, 100 Years of Music, Machines, and Money*. New York: Da Capo Press.
Copsey, Rob. 2018. The Official Top 20 Best-Selling Cassettes of 2018 so Far. *Official Charts*, July 25. http://www.officialcharts.com/chart-news/the-official-top-20-best-selling-cassettes-of-2018-so-far__23407/
Denegri-Knott, Janice, and Mike Molesworth. 2010. Concepts and Practices of Digital Virtual Consumption. *Consumption Markets and Culture* 13 (2): 109–132.
Duval, Luc. 2011. Fabricated Nostalgia: A Discussion of Vinyl Records. *The Pedagogic Verses*, January 9. http://pedagogic-verses.blogspot.com.au/2011/01/fabricated-nostalgia-discussion-of.html
Fitzpatrick, Kathleen Fitzpatrick. 2006. *The Anxiety of Obsolescence: The American Novel in the Age of Television*. Nashville, TN: Vanderbilt University Press.
Gandevia, Simon, and Uwe Proske. 2016. Proprioception: The Sense Within. *The Scientist*, September 1. https://www.the-scientist.com/features/proprioception-the-sense-within-32940
Glitsos, Laura. 2017. Ways of Feeling: The Transformation of Emotional Experience in Music Listening in the Context of Digitization. PhD diss., Curtin University.
Green, Ben. 2017. Having the Sceptre: Wu-Tang Clan and the Aura of Music in the Age of Digital Reproduction. *Popular Music* 36 (3): 427–440.
Greenring, Tanner, and Jack Shepherd. 2013. 35 Music Experiences You'll Never Have Again. *Buzzfeed*, August 16.
Griffith, James, and Jon Brown. 2014. *Cuban Fury*. Directed by James Griffith. London: StudioCanal.
Grosz, Elizabeth. 1994. *Volatile Bodies. Toward a Corporeal Feminism*. St. Leonards: Allen and Unwin.
Gunn, James. 2017. *Guardians of the Galaxy Volume 2*. Directed by James Gunn. Hollywood: Walt Disney Studios Motion Pictures.
Hansen, Mark. 2006. *Bodies in Code*. New York: Routledge.
Harrison, Anthony Kwame. 2006. Cheaper Than A CD, Plus We Really Mean It: Bay Area Underground Hip Hop Tapes as Subcultural Artefacts. *Popular Music* 2: 283–301.
Hornby, Nick. 1996. *High Fidelity*. London: Indigo.

Horton, Alex, and Wesley Lowery. 2018. 'Gun' Man Held Before Police Shot Him Dead in Backyard Was An iPhone. *The Sydney Morning Herald*, March 23. https://www.smh.com.au/world/north-america/gun-man-held-before-police-shot-him-dead-in-backyard-was-an-iphone-20180323-p4z5uw.html

Howes, David. 2005. *Empire of the Senses: The Sensual Culture Reader*. Oxford: Berg.

———. 2006. Scent, Sound and Synaesthesia: Intersensoriality and Material Culture Theory. In *Handbook of Material Culture*, ed. Chris Tilley, Webb Keane, Susanne Küchler, Mike Rowlands, and Patricia Spyer, 161–172. London: Sage.

Huizinga, Johan. 1955. *Homo Ludens: A Student of the Play Element in Culture*. Boston: Beacon Press.

Kafka, Franz. [1919] 1948. *The Penal Colony: Stories and Short Pieces*. New York: Schocken Books.

Kaplan, Louise. 2006. *Cultures of Fetishism*. New York: Palgrave Macmillan.

Kenny, Niall. 2013. *You Need To Hear This*. Noisey Video, 6:06. http://noisey.vice.com/en_uk/you-need-to-hear-this/a-short-film-about-cassettes

Laplanche, Jean, and Jean-Bertrand Pontalis. 1973. *The Language of Psychoanalysis*. New York: Norton.

Lief, Jason. 2014. A Material World: The Smell of Vinyl. *The Twelve*, December 19. http://blog.perspectivesjournal.org/2014/12/19/a-material-world-the-smell-of-vinyl/

Lunning, French. 2013. *Fetish Style*. London: Bloomsbury.

McCracken, Grant. 1988. *Culture and Consumption: New Approaches to the Symbolic Character of Consumer Goods and Activities*. Bloomington: Indiana University Press.

Milner, Greg. 2009. *Perfecting Sound Forever: The Story of Recorded*. London: Granta Books.

Nora, Pierre. 1989. Between Memory and History: Les Lieux de Memoire. *Representations* 26: 7–24.

Pitts, Victoria. 2003. *In the Flesh: The Cultural Politics of Body Modification*. New York: Palgrave Macmillan.

Pohlmann, Ken. 1989. *The Compact Disc: A Handbook of Theory and Use*. Madison, WI: A-R Editions.

Seed, David. 1985. The Narrative Method of Dracula. *Nineteenth-Century Fiction* 40 (1): 61–75. https://doi.org/10.2307/3044836.

Shuker, Roy. 2010. *Wax Trash and Vinyl Treasures: Record Collecting as a Social Practice*. Farnham: Ashgate.

Shuker, Roy. 2012. *Popular Music Culture: The Key Concepts*. 3rd ed. New York: Routledge.

Sterne, Jonathan. 2012. *MP3: The Meaning of a Format*. Durham, NC: Duke University Press.

Stratton, Jon. 2001. *The Desirable Body: Cultural Fetishism and the Erotics of Consumption*. Urbana: University of Illinois Press.
Strong, Catherine. 2015. Shaping The Past of Popular Music: Memory, Forgetting and Documenting. In *The Sage Handbook of Popular Music*, ed. Andy Bennett and Steve Waksman, 418–433. London: Sage.
Sullivan, Nikki. 2012. The Somatechnics of Perception and the Matter of the Non/Human: A Critical Response to the New Materialism. *European Journal of Women's Studies* 19 (3): 299–313.
Sweetman, Simon. 2011. Music Rules and Rituals. *Stuff*, June 20. http://www.stuff.co.nz/entertainment/blogs/blog-on-the-tracks/5164717/Music-rules-and-rituals
Taylor, Zack, and Georg Petzold. 2012. *Cassette: A Documentary Mixtape*. Cassette, 1:33. http://www.cassettefilm.com/#trailer
Tschmuck, Peter. 2006. *Creativity and Innovation in the Music Industry*. Dordrecht: Springer Netherlands.
U.S. Legal. 2001. Possession is Nine Points of the Law and Legal Definition. http://definitions.uslegal.com/p/possession-is-nine-points-of-the-law/
WeAreEvolving. 2018. Anyone Else Like the Smell of Old Albums. *Reddit.com*. https://www.reddit.com/r/vinyl/comments/7wggpa/anyone_else_like_the_smell_of_old_albums/
Weiss, Gail. 1999. *Body Images: Embodiment as Intercorporeality*. New York: Routledge.
Wolf, Charles. n.d. Proprioception. *Artbrain*. http://www.artbrain.org/proprioception/

CHAPTER 3

Liveness in the Age of Digitization

Somatechnics and Live Music

The thesis in this chapter calls for a rendering of the bodymind as fluid with the techniques through which the live music experience becomes legible, coherent, and pleasurable. I examine the ways in which bodily aspects, *soma*, play with, and are played upon by *tekné*, that is, the technologies, techniques and configurations of liveness which continue to reshape what it means, and how it feels, to experience live music.[1] It is fruitful here to recall the bodymind itself as an artefact of *somatekné*, the notion explored in the Introductory Chapter, which in this context suggests the bodymind as a crafted whole which constitutes itself as an experience of the live through techniques as they are inhered into the subjective and phenomenal. Simultaneously, we can think of the body as the material aspect of the subject that is consciously embedded in the live concert space and thus which *contributes* to its shape and quality.

This chapter takes Philip Auslander's work in *Liveness: Performance in a Mediatized Culture* (Auslander 1999) as a launching point. Auslander asserts that following the incorporation of video technology into the live model, we have witnessed "the precedence of the mediatized over the

[1] From here, I will refer to the 'live music event' or 'live music experience' as an umbrella term which refers to a range of live music design (the outdoor arena show, the smaller gig, the indoor concert and so forth), unless I differentiate in order to specify a particular point that is dependent on the physical configuration of the live event.

L. Glitsos, *Somatechnics and Popular Music in Digital Contexts*, Pop Music, Culture and Identity,
https://doi.org/10.1007/978-3-030-18122-2_3

live" (10). However, I extend and reconfigure this argument somewhat in order to suggest we are witnessing the *synthesis* of the mediatized with the live in new and unexpected ways that reshape the psycho-spatial boundary of the concert experience. Auslander in fact acknowledges and recognizes this tension—between the competing insistences between liveness and mediatization—in which the binary tradition seems to give way to a new, more complex organization. He ponders on, "The progressive diminution of previous distinctions between the live and the mediatized, in which live events are becoming more and more like mediatized ones" which he says raises the question "of whether there really are clear-cut ontological distinctions between live forms and mediatized ones" (7). He continues on to note that though his "initial arguments may seem to rest on the assumption that there are" ultimately he finds "that not to be the case" (7). It will be these latter assertions from which this argument takes its lead in order to argue that new media has in fact generated a paradoxical moment in which the singularity of liveness continues to be a salient aspect of music listening practice *despite* the complete assimilation of the mediatic with the live.

I trace this argument through several fundamental structures in which liveness might be considered as the primary aspect of pleasure, each revolving around time-boundedness and the body. To emphasize the relationship between the live music experience with the construction of time-boundedness, I begin with a discussion of Bernard Stiegler's *The Technics of Time* (1998). After which, I move to a discussion of the ways we might think of temporality in relation to the practice of dance and the ways in which dance can be read as a technic of listening pleasure. Moving on from a discussion of time-boundedness and temporality, I then initiate some thoughts on the ways in which the technics of vocality (both the fleshiness of the vocal folds as well as the hard technologies on which live music vocals rely) serve to differentiate some aspects between liveness and mediatization. The latter part of the chapter is concerned with unpacking the recent and ubiquitous integration of camera phone technology into the live concert space. I first situate the camera phone in the scope of Derrida's teletechnology to understand it as a device which helps produce a sense of presence or 'actuality,' to which I incorporate Merleau-Ponty. This approach situates the device as an extension of the body which pulls the absent social body into the individualized experience of the live. All of these discussions ultimately point to the paradoxical nature of the live versus mediatized relationship.

Stiegler's Technics of Time: Constructing the Live in the Field of the Now

The experience of listening to live music is differentiated from other music listening practices, such as listening to a compact disc for example, because of its ephemerality and thus its relationship to the construct of time. This, in itself, is a technique which renders the live space as a legible experience, that is, a way to make meaningful the listening experience as a unique and specific kind of practice. As I mention above, to begin, I draw from Bernard Stiegler's *Technics and Time* (1998) in order make visible one of the fundamental techniques that produces the pleasure of the live—the function of time; and moreover its construction as a direction, a flow and as an irreversibility. 'Somatechnically speaking,' we might say that we bring the technics of time into the body by way of technology—or that the body is technologized by the technics of time.

In the Preface to this work, Stiegler "apprehends" technics as the "horizon of all possibility to come and of all possibility of a future" (1998, ix). I use this as a premise of reading temporality as a distinguishing function of the live because the horizon of possibility marks the moment that separates the experience of the Now from the idea of the Future. To move this into a discussion of the live music experience, the 'horizon of all possibility' flavours live music listening experience with a magnitude of potential futures; what might happen in the show, how the tracks might be played differently, the feel of the crowd, the quality of the 'sound,' how the act performs and, in its totality, how the show unfolds as it comes together in the singular stream of time. The coalescing of the expected with the unexpected constructs the experience of the live concert because of this 'horizon of possibility.'

However, reading temporality as a distinguishing function of the live is not necessarily a pre-ordained construct, rather, it is produced by a technic of time. For Stiegler, "technics" is a "process of exteriorization [that marks] the pursuit of life by means other than life" (1998, 3), that is, the production and use of inorganic technical objects produces the quality and characteristics of certain experiences of time and its relationship to space. In another sense, technics are the modes and models which bring time into being, or as Stiegler writes, "As production (*poiēsis*), technics is a 'way of revealing.' Like *poiēsis*, it brings into being what is not" (9). In regards to music and the live music experience, we can turn to Stiegler's use of Aristotle to understand the quality of this *poiēsis*.

> In the *Nicomachean Ethics* Aristotle writes: Every art [*tekhnē*] is concerned with bringing something into being, and looks for technical and theoretical means of producing a thing which belongs to the category of possibility and the cause of which lies in the producer and not in what is produced. (Aristotle 1984, quoted in Stiegler 1998, 9)

The shape and quality, or what the audience might call the 'feel,' of a live music experience is the outcome of the means by which that experience is brought into being. This includes the narrative structure of the live experience, that is, the choice in which the set (the songs) are played, the intensity of playing itself and other mechanisms which produce tension in narrative form. Additionally, there are the hard technologies that produce the live music event in space such as the production equipment, as well as specialty skillsets including the application of live sound engineering. These are what Stiegler would call the "bringing-forth that is particular to technics" and which "constitutes a mode of truth" (1998, 9). For example, the production of lighting, pyrotechnics and particular sound schemas which also produce narrative tension which builds towards a culmination or 'climax' of the show. These all constitute a sense of temporality, of moving through 'time.' The conceptual 'mode of truth' returns here to the idea of a coalescing of the expected with the unexpected. Put otherwise, this might be understood as the coalescing of the Now with the 'horizon of future possibilities' which generates the ideal model of the live in which the experience takes place.

The conceptual 'mode of truth' also resonates with the discourses of authenticity that serve to construct the pleasure of the live music listening experience (Barker and Taylor 2007). We might also understand this process as a kind of *tekné*. As noted by Hugh Barker and Yuval Taylor, there are a range of techniques through which authenticity may be written, and consuming music in a live setting serves as a function of the meaning-making process and thus is implicated in shaping listening pleasure (2007, x). For example, live music constructs an ostensible link between the audience and the musician that feels as though it can only occur in *real time*, that is, in the 'authentic' setting of the live experience. In Wendy Fonarow's words:

> The audience wants to believe the musicians feel the emotion of the music, feel the performance, despite the fact that this may be the hundredth time the band has played this song in the last three months … In performance, the … audience wants to believe that what they are seeing is the *truth in music*. (2013, 203; original emphasis)

In another sense, there is an impression that in an 'authentic performance' truly 'anything might happen,' that in one moment, stability might give way to disaster, or that banality might give way to magnificence:

> Broken strings, broken instruments, or musicians falling over are not considered to be impediments unless they disrupt the performance. In fact, audience members often view troubles that performers need to overcome as enhancing the gig. One of the greatest differences in perceived quality of performance between musicians and audiences concerns technical difficulties. Often musicians, thinking of the minutiae of performance, feel that a show fraught with difficulty is a failure, while many audience members may consider it a triumph. What might be considered a mistake by the musician or an outsider— the wrong chord, or an instrument out of tune— are generally viewed by the audience as unique components of an original live event. (Fonarow 2013, 204)

The boundary of the live concert is just as much temporal as it is spatial because only the function of time can create the sense of anticipation through which techniques of authenticity are constructed in the live experience. Live music is thus exemplary of the use and application of technics in that its existence emerges from the matrix of production that brings it into being. The wonder of live music therefore lies in its production, not only in what is produced.

Dance and Disintegration in the Land of the Now

In this section, I argue that dance can serve as a technic of music listening. Dance lends itself to thinking through somatechnics in relation to temporality because dance is a technology of bodily configuration that occurs in set durations and often in relationship to music. In Paul Mason's words, "In time and space, dance and music unfold in parallel but independent planes that coalesce in mutually impacting cadences and segmentations" (2012, 6). However, Susan Leigh Foster reminds us that dance and music both complement as well as contrast each other, and that "These discrepancies point up the capacity found in both dance and music to articulate temporal forms while still demonstrating the autonomy of the two disciplines" (1986, 222). Dance is therefore a technic of listening pleasure because it is a process of coming-togethers, of divergences, of continuities and discontinuities of body parts, and ruptures of bodymind that occur *in a relationship to* music and thus brings to the music listening experience a variety of qualities for both the dancer and the audience.

It is worth returning to the discussion of *tekné* which I began in the Introductory Chapter to this book. The reader will recall that the neologism of 'somatechnics' sews together the "Greek *sôma* (body) and *τέχνη* (craftsmanship)" which, as Sullivan and Murray point out, "supplants the logic of the 'and,' suggesting that technés are not something we add or apply to the body, nor are they tools the embodied self employs to its own ends" (Sullivan and Murray 2009, 3). The authors are saying here that technics are what produce the body in the world. The authors emphasize that while somatechnics is useful in theorizing the coming together of bodies and technologies, moreover, it provides an ethico-political moment to understand the way in which the body is a craft that enacts itself as techné (in processes that form and de-form the body). The facticity of the body cannot be pulled apart from its functionality as a political and discursive construction which is produced by and with that body in culture. This is what is meant by 'technologies are not something added to bodies' (at least, not necessarily). This is not to say either that the body *is* technology, because that would suggest an *a priori* construction and thus negate the ways in which the body *becomes* an ethico-political artefact as it moves through the world. Rather, as Sullivan and Murray continue, we might think of "technés [as] the dynamic means in and through which corporealities and crafted, that is, continuously engendered in relation to others and to a world" (Sullivan and Murray 2009, 3). With this discussion replayed, I wish to talk about the ways in which dance, in the context of live popular music events, invokes techniques which produce the body-mind as concurrent with listening pleasure. Put more simply, we can read dance as a technic of listening pleasure.

In the first instance, I take dance as a technic of collective catharsis. Jenny Sundén reminds us that somatechnics (as a theoretical tool) takes the Spinozian affect model as its premise, and that, "a Spinozian understanding of the body as a relation is a fundamental opening up of, not only the body as system, but of its boundaries to the world and to other bodies" (Sundén 2013, 370). For example, dance is an 'opening up of the body to the world and to other bodies' in various ways. Let us take the instance of the contemporary dance most often witnessed at rock music festivals or metal music concerts: the mosh pit (sometimes referred to as a slam dance, referring to the way in which bodies 'slam' into each other) (Riches 2011). A mosh pit can be described as a group of individuals moving, jumping and 'head banging' in such close proximity to each other that they often bounce of one another, sometimes with such force that members of the mosh pit can be physically injured (Janchar et al. 2000). Particularly in hard rock and metal concerts, "Moshing is a ritualized and furious form of dancing which combines physical aggression with collective displays of emotion" (Riches 2011, 315). Bodies operate in such proximity in this setting that an individual can often throw their body on top of the mosh pit, or crowd, and be passed along without touching the ground until they are let down or, more often, are dropped (this is known as crowd surfing).

The mosh pit is overtly social; it blends together sweat, skin, flesh, smells, bodily discharges, energies, hair, spit, torsos in a violent yet joyous coming-together of human involvement with the music that is unfolding in space and time. Gabrielle Riches describes the mosh pit as a "vital part of the concert experience, providing an opportunity for metal fans to play with darker aspects of existence, subvert normative social conventions, and release pent-up frustrations of mundanity while fostering a strong sense of community" (316). Typically, the live music event is the only place this dance model can take place because it requires a certain number of (willing) bodies, a certain kind of music, and a particular set of circumstances (even as simple as whether or not a particular venue might allow mosh pits). It is a technology of bodily encounters that, in quite a visceral sense, 'opens up' the individual body to the social body, or following the Spinozian tradition, "opens up boundaries to the world and to other bodies" (Sundén 2013, 370). The somatic manifests the music as dance for expression as a social experience which reshapes its listening pleasure.

To take this further, the mosh pit as a somatechnique of listening experience can be often cathartic and also politically charged. Kelly Tatro's work (2014) gives us insight into the "hard-working bodies" of Mexico's punk scene. Tatro explains that, "Through physical exertion" punk musicians create an environment "in which fans may 'take out the rage,' producing solidarity through collective catharsis" (431). The intensity of this kind of dance functions as its affective 'fuel;' the intensity drives forward the motion of bodies smashing into one another and thus shapes the quality of the mosh pit—the more all-consuming the mosh pit then the more the dancing bodies might experience catharsis (Riches 2011, 326). However, as Tatro points out, this is catharsis with purpose and direction, and is social for a reason. By putting their bodies into this 'hard work' of moshing and colliding together, the punk scene "explores alternative notions of labor and value, attempting to enact their anarchist ideals in the context of the post-industrial, neoliberal metropolis" (431). Dance becomes a technology of affective-politico disruption which channels and directs the live music into the bodies of the scene members in very physical ways. Tatro continues on to note that:

> the powerful energy and intense distortion of the band's sound contributes to its affective impact, which inspires audiences to 'sacar la rabia' [*take out the rage*] in the performance space [which is] a means of describing a sense of catharsis that they enjoy at punk shows, one that provides them with a chance to express personal and collective rage through specific practices, such as screaming along with vocalists or participating in the slam dance. (Tatro 2014, 432)

I would emphasize here that this kind of focus on physicality of the live show is a significant point in the midst of a shift to digital technologies because through this example we see that physical techniques of music listening remain highly valued in the context of the new media environment. In fact, in David Hesmondhalgh's *Why Music Matters* (2013), he turns to dance as a way of defending the importance of music to the social body, as well as the individual body (109–111). I will not reproduce that discussion here, but I will use it as a vantage point in order to continue this discussion of the ways in which dance might be a considered a technology of movement that might shift, change, manifest, and queer the music listening experience.

Dance is a technic that interfaces the body with music, and by doing so accentuates the listening pleasure or, at other times, helps to disintegrate and/or queer the psycho-surreal boundaries between the music and the body. In this way, dance is a technic that inheres music into the somatic. Dance is a means by which we listen with our 'whole body.' Paul Mason's work on choreomusicology, which is "the study of the relationship between sound and movement within any performance genre" (2012, 6), gives us another insight into the way listening experience might be shaped by way of the technic of dance. For Mason, dance can provide music listening with "evocative inspiration" because there is a "dynamic reciprocity between both activities" (6). This may partly be because, as Mason writes:

> Organic forms of music are always a result of movement. Whether it is the slap of the chest, a clap of the hands or a stamp of the feet, sound arises from muscular effort. Even song 'is no more than a product of movement made

> by the various parts of the vocal tract reaching from the diaphragm to the lips' (Mithen 2005, 15). Music, as Blacking (1973) more deeply described it, 'begins as a stirring of the body' (111). (Mason 2012, 6)

To think of the relationship between music and dance in such a way suggests that dance is a vital aspect of a particular and singular way of music listening—a whole bodymind listening. Thus, dance is a somatechnique of music listening which occupies a special place in popular culture because dance operates as a technology of coming-togethers and potential disruptions between not only bodies and music but bodies and other bodies.

Taking this idea further, as a technology of political expression, dancing might be said to queer listening experience. I frame this argument through a parallel argument that Samantha Murray sets up in her work on the ways in which non-normative bodies disturb ideas around normative bodies. Murray writes about the way "that the fat body, in disturbing our ideals about normative embodiment and bodily aesthetics, is always already a *queer* body, or at least, that it is a body that queers" (2009, 153). Thus, in a similar vein, the ways in which dance suggests new modes of corporeal interpretation can lead to disruptions and destabilizations of normative corporeality, i.e. *dance is movement that queers.* For instance, take postmodern interpretive or experimental dance, that is, dance which interprets and plays with the relationship between sound and movement, between gesture and meaning. I am reminded here of the work of Marie Chouinard, in particular, the Sadler's Wells performance of bODY_rEMIX/ gOLDBERG_vARIATIONS (Jennings 2010) in which:

> dancers have been transformed by the addition of crutches, walking frames and prostheses. Naked except for a typically Chouinardian assortment of S&M harnesses, nipple caps and pointe shoes, they limp, swing and scoot across the stage to hashed-up Bach and a manipulated recording of a spoken commentary by the pianist Glenn Gould … A woman with miniature crutches drags herself about like a fish, gagging on a microphone; two women sharing a single leg brace perform a diagonal of mutant balletic leg lifts; a couple copulate while swinging from a harness. (Jennings 2010)

For Chouinard, "The work's movements plumb the insoluble mystery of the body, of the living being" (Jennings 2010). In dance, quite rightly we often assume that 'music does something to the body,' it moves us or perhaps, makes us move. However, here, I am saying that the body does something to the music. Dancing reconstitutes the music listening experience into a radically new dynamic that takes up space, or moreover, listening takes itself up as a body. Using Neitzsche, Philipa Rothfield suggests that "the body reproduces itself through dancing" (2009, 208). Dancing is thus a technology of musical interpretation and corporeal reconstitution; through certain gestures (and not without much skill and training) the body breaks itself down and rebuilds itself using music as its temporal parallel.[2] By doing so, both the performer of the dance, and those watching or engaging, are invited to reimagine different ways of bodily-being and bodily-doing which might suggest various incarnations of sexual and corporeal figurations.

The dancing body is a moving body, a moving machine, in both a theoretical and material sense. For example, as Karin Sellberg, Kamillea Aghtan and Lena Wånggren remark, the moving body is:

> a concept that bridges the divides between living and dead, solid and abstract, organic and inorganic substances—and as such it opens up the potential for numerous divergences with recent theories of embodiment and materiality. (2014, 1)

In this conceptualisation, the dancing/moving body has the ability to breach normative boundaries. It challenges those static and restrictive environments, environments which are constructed and maintained discursively and physically, and that keep at bay the potential for change to emerge—the office worker in her chair, the toll booth operator in his booth, and all those other sluggish (death-like) encounters which are naturalised by bureaucratic systems of power. Dancing threatens stillness and by doing so, threatens to change what *is*. As Michel Serres writes: "Dancing, the music of the body … sows the eternal return of rhythm with the seeds of the unexpected" (2008, 321). This is because change itself is movement, i.e., the passing of one state to another requires movement, just like dancing requires movement and therefore requires the body, and other bodies, to change. Lines of restriction, borders, *between bodies*, either human or nonhuman, may become redrawn and retraced.

[2] It should be noted that some postmodern dance can be performed without any music at all, but this is quite rare.

Somatechnics and Extreme Vocality: The Voice, Vibration and Liveness

In this section, I focus on the ways in which vocalists use the somatechnology of the human vocal folds in order to direct and mobilise certain listening pleasures in the live space. The emphasis on immediacy and temporality in the live experience also serves a vital function in the pleasure of listening in relation to the performance of vocality. Phenomenologically, this emphasis also serves to mark liveness as a differentiated listening practice from digital formations of pre-recorded vocal production. The most common assertion is that the production of overtly and overly produced vocals, particularly with autotune, tend to distance the listener and performer (Sclafani 2009). Often, the argument reverts back to the importance of authenticity in music (as we explore above), that is, the listener wants to believe they are actively listening to 'real' vocals in the live arena, and thus genuinely sharing an experience with that performance/performer (Hughes 2015). This is despite the fact that autotune can also be applied to the voice *in situ*. Regardless, historically, a critical component of appreciating live popular music is the experiential connection with vocality as it unfolds in the shared space of the performance. Nick Prior puts forth a compelling argument on this point drawing from Roland Barthes' notion of the "grain of the voice" from which Prior notes that pleasure emerges from the connection with the "form and physicality of the voice's originator—the body" and that, "the voice is at its best, for Barthes, when it is a physical expression of the membranes, the muscles and mucous of the body, when it is a performance of the flesh" (2018, 122). Again, this also points to a paradoxical construct emerging from the new kinds of syntheses between liveness and mediatization that thread throughout the chapter in the sense that constructing the live depends upon its relationship to the digital, recognizing each not as opposites but dependent variables. Much of popular music, as opposed to baroque or classical as just two examples, is heavily reliant on the performance of vocal lines, so an examination of the way in which vocality functions in the live space is extended here.

The points I make here are not to suggest that similar, or even the same, kind of connections might *not* be cultivated in listening to recorded music. Our reactions to music are dependent on a range of factors constantly shifting and moving at any given time and far too lengthy to list here. This discussion is also not to imply that one experience (listening to recorded music or listening live music) is superior. However, it is to emphasize that the specificity of the live event can produce or accentuate quite different affective encounters because the performer and listener share material space and share the physical interconnections which might be formed through that physicality.

One of the qualities I focus on that is most apparent in the live experience, as opposed to other non-live listening practices, is *loudness*. I take loudness is two ways. The first is the quality of loudness in relation to the decibel level which can be raised to suprahuman levels with the aid of amplification technologies. The second quality of loudness focuses on the production of sound through the performer's vocal folds which occurs through the play of fullness and silence using vibratory capacity. In order to make this argument in the most impactful way possible, I focus on the somatechnic of *extreme* vocality, through both its production (by the vocalist) and impact (on the listener) in the live space, which are brought together through technologies of sound (amplification).

Extreme vocality can be understood as the application of what Michael Edgerton terms as the "extra normal voice" (2014, 109). The extra-normal voice here relates to a vocal sonority and vocal performance that features atypical character relating (but not limited to) tessitura, timbre and rhythmicity, and textural qualities such as fry, grit or nasality, which sit outside mainstream practices of popular music singing styles within a specific cultural context, and which seek quite deliberately to produce atypical affective responses for the listener through the deployment of specific vocal structures. Various deployments of extra normal voice are what we can call the *somatekné* that produce extreme vocality.

To bring this examination to light, I focus for a moment on live vocal performances by the artist Diamanda Galás as a case study. Though Galás produced popular music on the fringes of mainstream listening culture, she was originally trained in the bel canto operatic tradition and became world-renown for what has been described as "sparse soundscapes of tormenting screams, incantations and poetry" which "showcased her famed vocal acrobatics" (Brown 2016). Galás routinely referenced political and social issues in her lyrics, with particular emphasis on mental health and feminism. That is to say that Galás used her vocality as a technic for political and affective expression. Most urgently for Galás in the 1980s, was her work calling for the awareness of the devastating effects of the AIDS virus on the gay community in San Francisco, a community which battled both the ravages and stigma of the condition, and which claimed the life of her brother, Philip-Dmitri Galás (Brown 2016).

One of the significant aspects of Galás' live vocal is 'nasality' and, in a moment, I will tie the experiential impacts of nasality to signifying practices of sound. In order to achieve hypernasality, certain affordances of the vocal folds are deployed in what I would call a somatechnique of bodily dynamic. Nasality, or what is often referred to as 'the glottal' sound, is generally produced through lowering the soft palate at the back end of the throat and pushing air through the nose. Hypernasality can be read, historically, in relation to the vilification of nasal twang in black choral singing. Grant Olwage reminds us that, for the 'vocal purists' "nasal 'twangs' are serious faults in singing…to be understood and avoided by sedulous practices with the mirror" (2004, 215). By casting hypernasality as impure and "guttural," ethnomusicologists once cast black singing, and thus blackness, as impure and guttural. This stigma is a legacy which continues to permeate vocal study today and which also gives us insight into the ways in which particular bodily techniques might be utilised to produce certain affective, perhaps only near-to-conscious, connections between singer and listener which are framed by social constructs and the racialization of what was (and can be) considered as 'good' and 'bad' human sounds.

Further, nasality was, and still is, also associated with class. Olwage points out that, the "hallmarks of the Cockney accent" such as "twangs and 'throaty enunciation' were the bodily symbols of a vocal sound that transgressed the idea of a grainless bourgeois voice" (2004, 215). The issue of race and class thus complicates a reading of the live vocal performance, and, in the case of Galás' hypernasality, her whiteness problematizes the deployment of the nasal tone. I would suggest that the use of nasality might be read as an appropriation of blackness in order to draw attention to the reproduction of all that has been disavowed by vocal pedagogy, and thus make visible the operations of race and class. This is not to conflate complex identification systems which position subjects through race or through sexual orientation, rather, it is to cognize ways in which vocal capacities mobilize as techné in order to sew together the discursive with the corporeal. For example, Olwage explains that this nasal tone was typically read as a "product of the 'chest register,'" which was considered as the "source and repository of sonic displeasures, such as the shout,"

> and against which voice culture fought; it 'must be silenced,' demanded the pedagogue John Spencer Curwen (1888 [1880], 315). The 'head voice,' advocated in its stead, produced a tone 'pure, melodious, resonant, refined,

> and blending', by giving the breath easy passage from the lungs/opening the throat, freeing the larynx/impelling it to the front of the mouth and out into the world, rather than trapping it within the body as guttural or nasal tone. (215)

I suggest here that Galás' use of the hypernasal tone is a somatechnique, that is, a skill of the 'vocal technology,' which is deployed in order to mobilise a sense of 'displeasure,' and perhaps 'discomfort' in the live arena. In terms of signifying practices, it arranges the confrontation with both sound and political complexity. Coupled with loudness in both senses of the term—in vibratory play and decibel limit—live vocals are differentiated from the listening experience of a recording that one can turn down or turn off, because it represents an embodied, collective and spatial-temporal coalescing between performer and listener. However, this kind of performance must be mediated in the first instance through the technology of sound production and, further, is often memorialized through the archive of video technology (or sometimes even simultaneously mediated through a screen in the concert space). This complicates the clear-cut suggestion that liveness is predicated on non-mediatic systems of production that emerge from physical in-space relationships. Again, we find ourselves back to the paradoxical nature of live music experience in that the entanglement between liveness and mediatization produces ontological slippages between our definitions of each.

Despite this paradoxical complexity, I wish to meditate further on the corporeality of the voice as a *vibratory technology*, or a somatechnique, and how the listener might be implicated in that somatechnique through the confrontation with extreme vocality in the concert space—in its loudness, quality and ambience. We can think about the vibratory technology of the voice using the example of Galás' again, and by calling forth Patricia Ticineto Clough's work on the physicality of sounds, particularly as they are transmitted through the somatechnic. In somewhat of a manifesto, Clough calls for "an ontology of vibrational force" in which that force "lets sound come to the rescue of thought rather than the other way around, forcing thought to vibrate, loosening up its organised or petrified body" (2013, 69). We might think of both the body of the vocalist and the listener as implicated in this 'loosening,' through what Clough calls a "resonating circuit of coming in and going out" (66). For example, I would take the string of live concerts that Galás performed in the 1980s during the Saint of the Pit tour (and the reader is directed to the view the live videos of which exist online, the irony of which points again to the paradoxes emerging in the play between the live and the mediatized) as a foundation from which to extend this theory.

In these performances, Galás deploys a combination of loudness (in the sense of a vocal play that hinges upon silence and fullness) with the technique of fry (a vocal quality), which is a raspiness or "creakiness" affected to the voice in which it appears as a forced or even painful whisper (McAlpine 2016).[3] However, in both techniques which produce loudness and fry, the vibratory function of the vocal folds are called forth in order to produce a wide range of affective fields. In the case of loudness coupled with fry, extreme vocality is calling attention to the vibratory capacity of the vocal technology, *particularly as a vibratory technology that works up on the body of the listener*. The listener's experience, I suggest, becomes an experience bound together in a material way through the vibratory field of resonance produced through the extreme vocal somatechnique. Douglas Kahn suggests that the physicality of loudness might be read "as the experience of the intensity of vibration on the whole body as well as within it" (1999, 227). Clough's work on the physicality of sounds alludes to a similar idea, in which she writes, "In the echoing of the other's voice, one feels the sonorous cave, the resonating drum that one becomes, that one is when sounding – speaking – singing" (2013, 66). The very fact that sound can burst an eardrum points to the significance of the ways in which sound, as vocal technology, interacts with the listener in absolutely material ways. The folds of the vocal apparatus affect the air in such a capacity, and are amplified to such a level, so as to produce resonance within the bodily register of the listener, thus affecting and shaping the affective limits of the live encounter and delineating the lines of pleasure within that encounter. In the context of new media and digitization, the significance of this process should be placed within the model of our understandings of liveness and mediatization.

However, it must be conceded that the loudness of the voice in the live arena can only *become* loudness through the affordance hard technologies (here I am referring to loudness in relation to decibels, as opposed to the use of vocality playing with fullness and silence, although the production of each are not entirely separate from the other). Without such amplification, extreme resonances produced through the vibratory technology of the vocal folds would not be anywhere near so affecting in the context of

[3] Although when the technique is executed correctly, it should not produce any pain for the vocalist.

the live performance (where performers are competing with spatialized structures, the noise of other voices, room dynamics, and in reference to open-air events, sheer size of crowds and arenas). The vocality then must be enhanced by sound production technology (technics, or what Stiegler calls the inorganic and exterior models of production), which work and act upon the voice.

The Technics of the Camera Phone in the Live Experience

Certainly one of the most noticeable shifts in the technological landscape of the live music space is the emergence of the smartphone. This is a shift which serves to further problematise the traditional binary structures which define liveness and mediatization. Jessa Lingel and Mor Naaman quite accurately point out that, "If the enduring image of concert-going in the 1960s was enthusiastic attendees waving their lighters in approval of an acoustic guitar set, in the 2000s, the prevalent view of live music could very well be a sea of music lovers with their mobile phones raised to capture video for rapid uploading to a variety of social media sites" (2012, 332). The fact that this technology has reshaped live music listening is incontestable, with research such as Lingel and Naaman's finding categorically that, "Practices of video recording during live music concerts have a direct effect on how music fans experience concerts" (338). However, I move away from Lingel and Naaman's research which tends to focus on the social media aspect of camera phone recording. I also distance the argument from the minutiae of the 'negative' or 'positive' aspects of this phenomenon. Rather I will explore the ways in which we might think through the utilization of the camera phone as a technic which reshapes our experience of the live music event through its absorption into bodily dynamic and as a product of the relationship between popular music with screen culture.

First, I wish to discuss the way that live music has formed a very specific relationship with mediatization in particular relation to screen culture, which can be thought through in terms of the Derridean theory of teletechnology, defined here as a technic that "produces, organises and mediates" time (Briggs 2015, 59). In simplest terms, the field of teletechnology refers to "the set of technologies, industries, professions, practices and conventions of what we more routinely call 'the media:' the press, television, cinema, radio, the Internet as well as related mobile, networked and digital channels or industries of public communication" (58). However, in more complex terms, teletechnology is a field designed to raise the "kinds of questions—aesthetic, ethical, political—that are raised by the fact of teletechnology" (59). By doing so, as Briggs notes, teletechnology serves to "displace" the more "family term 'media'" and with it our more "ordinary" and pre-figured understanding of these kinds of media technologies. It is for this reason that approaching the camera phone as a teletechnology helps to displace its more banal implications and excavate its mechanics as a somatechnic in the live arena.

At the most primary, the camera phone enacts a specific function of time in service of a live experience, to shape its boundaries and effects (the reader might remember 'time moving differently' when at a concert, some reporting that it seems to either stand still or move faster). For Derrida, digital media, more than ever before, has 'brought into play' the artificiality of time. During a set of filmed interviews, which were transposed into the printed work *Echographies of Television* (2002), Derrida spoke of the ways in which time and mediatization produce each other. For Derrida, much as in the vein of Steigler (who co-wrote the book), "[time] is an *artifact*. In its very happening, the time of this public gesture is calculated, constrained, 'formatted,' 'initialized' by a media apparatus" (2002, 3). Robert Briggs' analysis of this work points us to the significance of this statement in relation to the construction of live music, in which he writes:

> Derrida's discussion of the media in terms of teletechnology thus serves a double purpose: on the one hand it links reflection on media technologies to a broader concept and history of technology generally, while on the other hand it orients deconstruction in this case not so much towards the philosophical or logocentric production of presence (in the form of ideal objects) as towards the technical and industrial construction of *presence* today in the form of mediatized 'actuality' or 'liveness.' (Briggs 2015, 59; original emphasis)

The production of *presence* and the construction of *liveness* as a mode of authenticity in the navigation of time is crucial in understanding the ways in which camera phone devices, as teletechnologies, serve the live music listening experience 'somatechnically' (that is, through a non-organic aspect of technical integration). I will return to this discussion shortly to unpack the ways in which this happens.

However, for a moment, I wish to pause this discussion in order to trace the co-emergence of liveness, music events, and teletechnology. In many ways live music, as opposed to pre-recorded music, has become bound to visual representation since the mid-twentieth century. Since at least the 1950s, live music (and especially the live concert event) has developed as a thing *to be filmed*. Vincent LoBrutto explains how in the early 1950s, "live broadcasts were the essence of television" (2018, 349). Together, live music and recording technologies have co-shaped processes of apprehension for what is often referred to as 'being in the moment' of live music. LoBrutto refers to popular culture critic Gilbert Seldes who, in 1952, wrote that, "'The tension that suffuses the atmosphere of a live production is a special thing to which audiences respond; they feel that what they see and hear is happening in the present and therefore more real than anything taken and cut and dried, which has the feel of the past'" (as cited in LoBrutto 2018, 349). This language emerges from the discourse of authenticity that works to construct the live experience in the wider discourse of music listening culture, which I begin to discuss above. Listening culture has become so profoundly entangled with the televisual apparatus—in particular the interface of the screen—that all music experience, including liveness, has become subsumed into the "mediatic system" (Jameson 1991, 162). This process is naturalised by the commercial music aesthetic which produces a subject who calls upon the screen interface in order to 'make meaning' of live music through visual codes.

From the late-1960s, large live music concerts and festivals became more and more mediated through the televisual apparatus (Hill 2006). This process concretized the tradition of experiencing the live through, and with, the mediatized. This is of particular issue because I suggest that this produces the context through which we later see the relationship between the camera phone and the concert emerge. We can trace an aspect of this process to the early live music documentaries of the 1960s, the first of which is *Monterey Pop* (1968) directed by Don Alan Pennebaker, which was based on the first Monterey Pop music festival in 1967. In her analysis of Otis Redding's performance in this documentary, Sarah Hill writes that, the audio recording:

> Captures the raw emotion of Redding's delivery, but the film captures the playfulness behind it. Pennebaker was very careful to edit footage taken from the rear of the stage for this song, providing an intimate account of the dialogue between the singer and Booker T. and the MCs, and also between

> the singer and the audience. Just as significant an effect is the interference of the lighting, alternately obliterating Redding from view and embracing him in a warm halo of light. It allowed the film audience to witness the union of Redding and his new community … suggestive of a personal motion toward communion. (Hill 2006, 36)

In this respect, the live performance is constructed through specific codes which privilege certain information, and thus configure, or 'write' the narrative through the process of recording it. Following on from this, we might say that live music is pre-figured by the act of recording. The live performance is *produced* through the processes of recording, editing, production and archiving, and thus, the live performance as a cultural artefact is entwined with the aspects of that production.

In a similar sense, Michael Wadleigh's 'rockumentary' *Woodstock* (1970), based on the 1969 Woodstock Music and Arts Festival, is another early enterprise in the construction of liveness and the development of the live experience as a thing to be filmed. What interests me here is not the construction of Woodstock through mediation *per se* (though this point is significant in its own right and well theorised in Andy Bennett's edited collection *Remembering Woodstock* [2004]). Rather, I am concerned with the way that the act of *filming* itself became concretized as a tradition, and how the function of filming live music became part of popular music culture in its own distinct way, that is, not the artefact of film but the process of filming. As Dave Laing explains, the *Woodstock* film was generally accepted to capture the *zeitgeist* of a live popular music event (2004, 13). The underlying notion carrying this assertion is that liveness is an event that *should* be filmed and that filming itself is an integral aspect of the live experience. Laing goes on to explain that the distinct difference of the Woodstock movie to prior unsuccessful attempts at mediating live concerts was the way the filming and editing were approached (13). Indeed, the promoters believed that "the solution to this dilemma was to hire a documentary film-maker who was as unorthodox as themselves" (13) and so Michael Wadleigh's succeeding film used "bold technical means" (13) to construct a sense of excitement and ambience. In particular, Wadleigh deployed split-screen editing (where two different images appear simultaneously) which would be described as "'a milestone in artistic collage of raw footage into a multipanel, variable-frame, dazzling montage that engages the sense with barely a let-up'" (Elley

1991, 682 quoted in Laing 2004, 13). What this does is inhere the aspects of filming, and processes of editing, into the paradigm of liveness in a way that sets up the concert as a thing *to be filmed*. Concerts have since become appropriated by audience members as a thing to be objectified; filmed, edited, split-panelled, re-represented and narrativized in filmic terms; demonstrated quite clearly by the ubiquity of camera phones at concerts. The ability to film the concert has become just as important, if not more important in some senses and for some fans, as experiencing that concert first-hand, which I argue is a product of the synthesis between live music and filming processes that emerged during this critical phase of popular music.

This brings us to an historical moment, the late-2010s, in which mobile camera phone technology serves to construct presence "in the form of mediatized 'actuality' or 'liveness'" (Briggs 2015, 59). In particular, this process constitutes the live music listening experience as a coherent practice for a demographic of fans for whom liveness and filming did not emerge but were *always-already* tied together. The way in which live music events became subsumed into the mediatic system has extended beyond that of the televisual apparatus and become inhered into the digital filmic practices of camera phone teletechnology.

With this foundation being laid, I return now to the discussion initiated above regarding the ways in which camera phone devices, as a teletechnology, serve the live music listening experience. For example, *how* do live music fans utilize the camera phone as a function that produces this 'presence' of actuality? In the first instance, I raise the notion of 'the music collector'. Usually, as Roy Shuker advises in *Wax Trash and Vinyl Treasures* (2010), "The popular image of contemporary record collectors is of obsessive males, whose 'trainspotting' passion for collecting is often a substitute for 'real' social relationships," which Shuker has termed "the High Fidelity stereotype" after the book by Nick Hornby (33). However, in my narrative, we might also think of the contemporary music collector as the hyper-digital, socially-connected 'millennial' whose cultural currency is built by the pleasures of collecting experiences. For decades, teletechnologies have been bound to practices around the 'collecting of moments' or in another sense, collecting that ephemeral and elusive artefact of *time*. By doing so, the teletechnology has become the somatechnic by which culture historicizes and memorializes, or we might say, it is a somatechnic through which one makes liveness sacred and part of the self. To follow Margrit Shildrick's work on somatechnics, understanding the symbiotic processes of human encounters with technologies can be thought through using a Deleuzian approach,

> in that what is of consequence is not the "content" of any particular entity—organic or nonorganic—but the provisional instantiations that emerge from the interconnectivity of multiple forces. There are no solid bodies as such, only becoming-bodies. As Deleuze and Guattari put it: 'tools only exist in relation to the interminglings they make possible or that make them possible'. The stirrup entails a new man–horse symbiosis that at the same time entails new weapons and new instruments. Tools are inseparable from symbioses or amalgamations defining a Nature–Society machinic assemblage.... a society is defined by its amalgamations, not by its tools. (Deleuze and Guattari 1987 cited in Shildrick 2015, 18)

Thus, I put forth the camera phone device as an amalgamation which has extended the pursuit of 'presence.' If the 'stirrup entails a new man–horse symbiosis that at the same time entails new weapons and new instruments,' then the camera phone entails a new *fan-show* symbiosis that at the same time entails new listening practices and new forms of pleasure wrought by the pursuit of presence.

Further, utilizing the camera phone in the concert space to collect experiences around live music is just as a much a somatechnic of listening as it is of self-narration, self-construction, and self-identification. Put in other terms, the camera phone technologizes identity-making mechanisms in new ways. In their work, George Ritzer, Paul Dean and Nathan Jurgenson elucidate the ways in which collecting experiences is of growing importance in contemporary Western culture (2012, 7). This is driven by effective marketing strategies of current digitech companies, in that every smartphone campaign is designed to convince the consumer that moments must be recorded in order to be concretised as 'real.' Although collecting photos and videos of concert experience is not a new phenomenon, its extent is unmatched in digital contexts. As a result, digital camera phones enable an unprecedented level of self-narration and the construction of emotional narratives—the stories we tell of ourselves to others, the stories we tell to ourselves about ourselves, and the emotional tone those narratives take. For example, in their article, "Tourist Photographs: Signs of Self," Russell Belk and Joyce Hsiu-yen Yeh's explain that, by taking photos we engage in a type of "self-fashioning" (2011, 349). The ubiquitousness of the camera phone in the concert space points to chronic and constant *collection* of self-fashioning material that can make sense of one's life and situate one's life within comfortable emotional conditions: photos attest to the participation in live concerts, of loving music, of being happy, of having friends, and enjoying one's life. As Belk and Hsiu-yen Yeh explain:

> the performativity of such photography with its staging and posing of shots means that tourists intend something more than simply experiential documentation. Along with the act of naming or labelling things we encounter, either before or after photographing them, in taking tourist photos we are potentially collecting illustrations and titles for a self-narrative. (2011, 349)

The photographer becomes both hero and narrator of their own epic and the practice of experience collection accumulates a wealth of resources from which to "extend the self" (Belk 1988). Belk and Hsiu-yen Yeh's write that, "By allowing us to collect evidence of where we have been and what we saw and did there, we may be attempting to claim these otherwise intangible and ephemeral experiences as a part of our extended self" (2011, 346). Digital filmic technologies have interpenetrated the concert experience and brought with them all of the somatechnics implied by collecting and preserving. As Laura Mulvey claims, "The technological drive towards photography and film had always been animated by the aspiration to preserve the fleeting instability of reality and the passing of time in a fixed image" (2006, 18). The coming together of camera phone technology with the concert experience produces a trend towards those screen relations that are driven by the synthesis of the live with the mediatized.

However, what is at stake in the symbiotic relationship between the human and the camera phone device is the growing tension between possessing the experience *or* taking part in that experience without the cultural prefiguration which assumes the experience as an artefact to be chronicled and preserved. In part, this tension is complicated by the notion that the desire to preserve and possess the ephemeral is not the same as actually *being able* to preserve and possess the ephemeral. We might draw a parallel to other forms of teletechnology. For example, in writing about cinema, Mulvey explains that "experience is so ephemeral, it has always been difficult to hold on to its precious moments, images and, most particularly, its idols" (2006, 161). In response to this ephemerality, the film industry provides still images of the movies that can supplement that movie-going experience (161). This is "designed to give the film fan the illusion of possession, making a bridge between the irretrievable spectacle and the individual's imagination" (161). Similarly, the concert experience is also ephemeral (in many ways much more so than the cinematic experience because a film cannot change upon each viewing but each live performance must change, or in another sense, a live performance cannot be guaranteed to be reproduced exactly). The camera phone therefore provides a similar function of the cinema stills. There is a sense that the obsessive chronicling of the event can somehow quash the anxiety of ephemerality and losing that experience in the march of time: the camera phone user becomes a kind of lepidopterist—pinning down the music experience behind 'glass' (or a screen, in the case of camera phones). Liveness is fleeting and so the camera phone works to hold that moment, suspended in time. What I mean by this metaphor is that by demanding the evidence of being in moment in order to 'save' the moment forever, many music fans (and many artists too) feel the action paradoxically destroys the thing it is trying to save (see Bennett 2012). For instance, in the same way that the glory of the butterfly is in its *livingness*, so too is the pleasure of the concert in its live-ness. The concert fan can preserve the representation of the experience but not the experience itself, affirming Mulvey's suggestion that these processes of cinema and photography are designed to give the fan the *illusion* of possession and, therefore, perhaps the illusion of control over the ephemerality of time itself. The camera phone is thus the technic of time-retrieval, a mastery of control over the uncontrollable—thus another a paradox which serves to further problematize notions of liveness and mediatized in the digital context.

Somatechnics of Social Listening: Pulling the Absent into the Present

Liveness is generated not only by the spatial-temporal boundary but by the insistence of listening pleasure as bound to the social and physical. Here I discuss the way that the camera phone technologizes listening pleasure in the live music space by way of socializing the body through new avenues. As a wide variety of theorists have suggested, the live music arena is a highly socialized space and in the context of popular music communities, it always has been (Hesmondhalgh 2013; Chesher 2007). Further, there are several insightful works which have begun to unpack the ways in which the camera phone enacts certain kinds of sociality, such as Chris Chesher's work on patterns of use at large concerts (2007), and Couldry's work on liveness and "the mediated habitus" (2004). Lucy Bennett's research also elucidates how the ideal of 'being there' has been redefined by social media communities interacting in real time (2012). Following on from Bennett, I move to suggest that the camera phone acts as a social-making mechanism that calls forth a kind of *absent-body*—the body of the missing spectator. By doing so, the camera phone inheres the social body *into* the individual body by way of the somatechnic of virtuality as offered through the device.

This process is ultimately naturalized in service of the concert experience in the post-2010s contemporary listening practice. Lucy Bennett's research demonstrates the way "online tools are involving individuals who are not physically present at the [live] show, seemingly incorporating them into the real-time 'live' experience" (2012, 545). Bennett traces certain strategies in which "fans use social and mobile media in an attempt to contest and reshape the boundaries of live music concerts, a practice that works to re-appropriate ideas of immersion in 'liveness'" (545). I take Bennett's approach as a launching point to suggest that these strategies not only reshape the boundaries of liveness, but also reshape what it means to be a body experiencing the live, both for the body-in-space and the absent-body. To do his I synthesize the work of Merleu-Ponty with Bennett, below.

For Merleau-Ponty, technics is that which extends bodily schemas. Mark Hansen expounds on Merleu-Ponty to explain the crux of this notion to tell us that:

> Whereas the body image characterizes and is generated from a primarily visual apprehension of the body as an external object, the body schema emerges from what, with autopoietic theory, we have called the operational perspective of the embodied organism. As such, it encompasses an 'originary,' preobjective process of world constitution that, by giving priority to the internal perspective of the organism, paradoxically includes what is outside its body proper. (Hansen 2006, 38–39)

As a result of this particular incarnation of bodily mapping, the 'body-in-the-world' becomes a site of pure potentialities, from which technics, Hansen argues, cannot be excluded (43). To unpack this point, Merleau-Ponty puts forth the analysis of the 'blind man's stick.' That is, the stick (as technic) which a blind man may use to navigate his way in the world, and locate the place of his body in that world,

> ceases to be an object for him, and is no longer perceived for itself: its point has become an area of sensitivity, extending the scope and active radius of touch, and providing a parallel to sight. ... To get used to a hat, a car or a stick is to be transplanted into them, or conversely, to convert them into the bulk of our own body. (1962, 143)

To transpose this to the somatechnic of camera phone technology, I read the device as the technic through which the individual body appropriates the social body, that is, 'converts' the social body 'into the bulk of the individual body'. For example, one present individual—through the prosthetic of the phone—brings into being an entire community within themselves. They are not only representative of a community; the individual embodies the absent body on their behalf. Following Bennett's argument, fans "contest and reshape the boundaries of live music concerts, a practice that works to re-appropriate ideas of immersion in 'liveness'" (545). For instance, we might think of the ways in which concert fans can call friends or family during a concert so the absent spectator can 'listen in on' the experience, or, more common now, is the way in which a concert fan might hold up their phone so that the absent body might 'look in on' the experience. The present body 'stands in' as a kind of conduit that, with the technic of the camera phone, is mediating the live experience beyond itself. Using Hansen's line of thought, the camera phone, like "the stick does not function as an explicit, cognitively assembled enhancement of the body image, but rather as an immediately practical, unthematizable expansion of the body schema" (2006, 43). The constant and perhaps chronic use of the camera phone as a way of experiencing of the live event adds to its technicity, that is, its normalization and integration into bodily schema works like the stick for the blind man. Merleau-Ponty suggests that it is *habit* that "expresses our power of dilating our being-in-the-world," which more and more means "changing our existence by appropriating fresh instruments" (1962, 143). The individual experience gives way to social encounter by way of operating under the means of the technic which renders the absent body *present*. The absent bodies become manifest, through all those channels which have been thoroughly naturalized by camera phone technology such as social media exchange and instant messaging services along with geotagging and advanced photographic technology. The relationship between liveness and mediatization again becomes re-articulated through a 'bringing-into-being' enabled by new screen culture, which not only extends the experience of the imagined fan community but re-invests the present body with that of the absent body.

Conclusion

This chapter has been an expedition in understanding listening experience and listening pleasure in the live space as it emerges within the post-millennium digital context. I have argued that a range of phenomena—from dancing, extreme vocality, and the use of teletechnology, take on very different qualities in the context of new media and the growth of digital spaces. To execute this argument, I have brought to bear an understanding of these practices through the theoretical basis of somatechnics; a basis which emphasizes the role of bodily configurations and techniques in the interactions brought about by new listening practices. By doing so, I also suggest a range of implications for the bodymind in relation to time-boundedness, temporality and the construction of time as contingent with live listening pleasure.

This chapter moves the discussion of popular music listening practices towards an appreciation of the holistic bodymind experience as indistinguishable from the production of both discursive and material constructs brought into being by technics.

References

Auslander, Philip. 1999. *Liveness: Performance in a Mediatized Culture*. Hoboken, NJ: Taylor & Francis.

Barker, Hugh, and Yuval Taylor. 2007. *Faking It: The Quest for Authenticity in Popular Music*. New York: W. W. Norton.

Belk, Russell. 1988. Possessions and the Extended Self. *Journal of Consumer Research* 15 (2): 139–168.

Belk, Russell, and Joyce Hsiu-yen Yeh. 2011. Tourist Photographs: Signs of Self. *International Journal of Culture, Tourism and Hospitality Research* 5 (4): 345–353.

Bennett, Andy. 2004. *Remembering Woodstock*. Aldershot: Ashgate.

Bennett, Lucy. 2012. Patterns of Listening Through Social Media: Online Fan Engagement with the Live Music Experience. *Social Semiotics* 22 (2): 545–557.

Briggs, R. 2015. Teletechnology. In *Jacques Derrida: Key Concepts*, ed. Claire Colebrook, 58–67. London: Routledge.

Brown, Louise. 2016. The Immortal Diamanda Galás: Still Wild, Still Extreme. *Noisy Vice*, March 28. https://noisey.vice.com/en_au/article/diamanda-galas-interview-2016

Chesher, Chris. 2007. Becoming the Milky Way: Mobile Phones and Actor Networks at a U2 Concert. *Continuum* 21 (2): 217–225.

Couldry, Nick. 2004. Liveness, 'Reality,' and the Mediated Habitus from Television to the Mobile Phone. *The Communication Review* 7 (4): 353–361.

Derrida, Jacques, and Bernard Stiegler. 2002. *Echographies of Television: Filmed Interviews*. Maiden, MA: Polity Press.

Edgerton, Michael. 2014. The Extra-Normal Voice in Singing. In *Teaching Singing in the 21st Century*, ed. Scott Harrison and Jessica O'Bryan, 109–132. Dordrecht: Springer.

Fonarow, Wendy. 2013. *Empire of Dirt: The Aesthetics and Rituals of British Indie Music*. Middletown, CT: Wesleyan University Press.

Hansen, Mark. 2006. *Bodies in Code: Interfaces with Digital Media*. New York: Routledge.

Hesmondhalgh, David. 2013. *Why Music Matters*. Chichester: Wiley-Blackwell.

Hill, Sarah. 2006. When Deep Soul Met the Love Crowd: Otis Redding, Monterey Pop Festival, June 17, 1967. In *Performance and Popular Music: History, Place, and Time*, ed. Ian Inglis, 28–40. Aldershot: Ashgate.

Hughes, Diane. 2015. Technological Pitch Correction: Controversy, Contexts, and Considerations. *Journal of Singing* 71 (5): 587–594.

Jameson, Frederic. 1991. *Postmodernism, or, the Cultural Logic of Late Capitalism*. London: Verso.

Janchar, Timothy, Samaddar Chris, and David Milzman. 2000. The Mosh Pit Experience: Emergency Medical Care for Concert Injuries. *American Journal of Emergency Medicine* 18 (1): 62–63.

Jennings, Luke. 2010. Body_Remix/Goldberg_Variations; Gustavia; Susan and Darren. *The Guardian*, May 16. https://www.theguardian.com/stage/2010/may/16/body-remix-goldberg-gustavia-susan-darren

Kahn, Douglas. 1999. *Noise, Water, Meat: A History of Sound in the Arts*. Cambridge, MA: MIT Press.

Laing, Dave. 2004. The Three Woodstocks and the Live Music Scene. In *Remembering Woodstock*, ed. Andy Bennet, 1–18. Aldershot: Ashgate.

Leigh Foster, Susan. 1986. *Reading Dancing: Bodies and Subjects in Contemporary American Dance*. Berkeley: University of California Press.

Lingel, Jessa, and Mor Naaman. 2012. You Should Have Been There, Man: Live Music, DIY Content and Online Communities. *New Media and Society* 14 (2): 332–349.

LoBrutto, Vincent. 2018. *TV in the USA: A History of Icons, Idols, and Ideas: Volume 1*. Santa Barbara, CA: Greenwood.

Mason, Paul. 2012. Music, Dance and the Total Art Work: Choreomusicology in Theory and Practice. *Research in Dance Education* 13 (1): 5–24.

McAlpine, Alissa. 2016. An Examination of Vocal Fry as a Feminine Identity Marker. *Academic Excellence Showcase Proceedings* 47. https://digitalcommons.wou.edu/aes/47

Merleau-Ponty, Maurice. 1962. *The Phenomenology of Perception.* London: Routledge.

Mulvey, Laura. 2006. *Death 24x a Second: Stillness and the Moving Image.* London: Reaktion Books.

Murray, Samantha. 2009. 'Banded Bodies:' The Somatechnics of Gastric Banding. In *Somatechnics: Queering the Technologisation of Bodies*, ed. Nikki Sullivan and Samantha Murray, 153–170. Farnham: Ashgate.

Olwage, Grant. 2004. The Class and Colour of Tone: An Essay on the Social History of Vocal Timbre. *Ethnomusicology Forum* 13 (2): 203–226.

Prior, Nick. 2018. *Popular Music, Digital Technology, and Society.* London: Sage.

Riches, Gabrielle. 2011. Embracing the Chaos: Mosh Pits, Extreme Metal Music and Liminality. *Journal for Cultural Research* 15 (3): 315–332.

Ritzer, George, Paul Dean, and Nathan Jurgenson. 2012. The Coming Age of the Prosumer. *American Behavioral Scientist* 56 (4): 379–398.

Rothfield, Philipa. 2009. 'Between the Foot and the Floor:' Dancing with Nietzsche and Klossowski. In *Somatechnics: Queering the Technologisation of Bodies*, ed. Nikki Sullivan and Samantha Murray, 227–224. Farnham: Ashgate.

Sclafani, Tony. 2009. 'Oh, My Ears! Auto-Tune is Ruining Music:' The Device Makes Everyone Sound Good, Which Isn't so Good. *MSNBC*, June 2. http://www.today.com/id/30969073/ns/today-today_entertainment/t/oh-my-ears-auto-tune-ruining-music/#.W3t_p5P-i_s

Serres, Michel. 2008. *The Five Senses: A Philosophy of Mingled Bodies.* London: Continuum.

Shildrick, Margrit. 2015. 'Why Should Our Bodies End at the Skin?:' Embodiment, Boundaries, and Somatechnics. *Hypatia* 30 (1): 13–29.

Shuker, Roy. 2010. *Wax Trash and Vinyl Treasures: Record Collecting as a Social Practice.* Farnham: Ashgate.

Stiegler, Bernard. 1998. *Technics and Time: The Fault of Epimetheus.* Stanford, CA: Stanford University Press.

Sullivan, Nikki, and Samantha Murray, eds. 2009. *Somatechnics: Queering the Technologisation of Bodies.* Farnham: Ashgate.

Sundén, Jenny. 2013. Corporeal Anachronisms: Notes on Affect, Relationality, and Power in Steampunk. *Somatechnics* 3 (2): 369–386.

Tatro, Kelley. 2014. The Hard Work of Screaming: Physical Exertion and Affective Labor Among Mexico City's Punk Vocalists. *Ethnomusicology* 58 (3): 431–453.

Ticineto Clough, Patricia. 2013. My Mother's Scream. In *Sound, Music, Affect: Theorizing Sonic Experience*, ed. Marie Thompson and Ian Biddle, 65–71. New York: Bloomsbury Academic.

Wadleigh, Michael. 1970. *Woodstock.* Warner Bros.

CHAPTER 4

Screen as Skin: The Somatechnics of Touchscreen Music Media

Introduction

In this chapter I explore the way mobile music devices with touchscreen technology produce new somatechnical figurations that reshape the emotional dynamics of music listening. Up until now, somatechnics has been largely applied in queer theory to bring forth and denaturalise "operations of power that shape corporealities" (Sullivan and Murray 2009). However, I extend its use here in order to argue that the changing relationships between the human-computer interface result in new affective schemas that expand and reconfigure how it feels to listen to music in a mobile setting. In particular, I focus on skin-on-screen contact in order to suggest that the screen acts as a reflexive surface producing intimate relations for the mobile listener. Touchscreens imply the relationship between skin on skin—the skin of our body (in particular the hands) against the skin of the screen. It follows that mobile touchscreen devices suggest a degree of sensuality—in the coming together of bodies, fluids and other organic materials which 'stick' to the touchscreen, the language of 'stickiness' pointing again to Sara Ahmed's conceptualization of the way affect can 'stick' to bodies, as discussed in the Introductory Chapter. Following the work of Ahmed and Stacey in *Thinking Through the Skin* (2003), I carry out a "dermography" of touchscreens, or, the study of surfaces as skin and skins as surfaces.

L. Glitsos, *Somatechnics and Popular Music in Digital Contexts*, Pop Music, Culture and Identity,
https://doi.org/10.1007/978-3-030-18122-2_4

The function of skin, both in a corporeal and a discursive sense, cannot be overstated. Skin is the covering that "protects us from others and exposes us to them" (Cataldi 1993, 145). Skin is profoundly significant in that it provides the basis for an overwhelming variety of trends in the politics of subjectivity, from the classic work of Frantz Fanon on skin colour (see *Black Skin, White Masks* [1952] 1967) to more recent feminist work on the politics of ageing skin or stretching skin, such as in the case of pregnancy (Tyler 2001). Skin is not politically benign. It is "the fleshy interface between bodies and worlds" (Ahmed and Stacey 2003, 7). By "thinking through the skin," to use Ahmed and Stacey's words, I read mobile touchscreen technology as an exciting new way to imagine music listening in terms of cyborgian relations.

In addition to performing a dermography of the skin-on-screen relation, I also put forth the mobile media device as a techno-*con*corporeal prostheticised figuration at the interface of human-computer relations. These devices are *con*corporeal, to borrow a term from Margrit Shildrick, because they can work prosthetically 'side-by-side,' as well as weaved within, the organic body (2009). In this respect, I focus on the somatechnics of mobile music players as sex toys (marketed as 'acsexories'). The somatechnics of this relationship, with particular emphasis on the technology of female genitalia, suggests a radical redefinition of how music shapes the body and in what ways bodies can be manipulated by the mobile music device. In this section, I interrogate the theoretical and material limits that demarcate the human and the machine in the scope of music listening. By doing so, I suggest that these prosthetic figurations are the basis for extant modes of inter- and intra- corporeal relationships with music devices, whereby music is absorbed into the subcutaneous layers of embodiment and therefore experience.

Mobile music media has become the subject of great interest in popular music studies in recent years (Werner 2015; Beer 2010; Åman and Liikkanen 2013). One of the most notable theorists is Michael Bull, who has focused on the "culture of iPod" in relation to the way sound maps out listeners' spatial awareness in the urban landscape and in the commuting experience (2000, 2005, 2008). Jonathan Sterne, too, has offered insights focusing on the socio-technological aspects of MP3 development to provide an understanding of the way that the "quality of 'portability' is central to the history of auditory representation" (2006, 825). However, as Nina Gram has pointed out, the literature focuses overwhelmingly on the motivations and effects of using mobile media in relation to exercising control over the listening experience in particular spaces (2009, 1).

I break from this trajectory to develop an inquiry that is based on the more recent developments of touchscreen technology in mobile music devices. Touchscreen technology is used in several portable media players such as the Pono, the Sony Walkman NW2-F805, the Cowon A5 Plenue and of course, the Apple iPod and iPad. There are two forms of touchscreen technology: resistive and capacitive. Both Android and Apple devices employ capacitive sensing whereby the screen is fitted with an electromagnetic field that can detect movement based on precise changes in that field from another conductive surface that transmits electricity—such as the skin (Kirk 2012). So, a touchscreen will only respond to something that emits electrical pulse, either through the living tissue of the skin or a purpose-built device that is designed to give an electrical pulse, such as a stylus that mimics living skin. The screen is therefore designed to respond to the *living-ness* of human touch.[1] In a way, the screen can feel our touch in the same way that we can feel the touch of others, again suggesting the touchscreen itself as a kind of skin. The touchscreen-skin therefore reproduces the human-computer interface as a reciprocal, affective, and deeply intimate phenomenon, in ways that we are only just beginning to experience in the development of portable music.

Somatechnics as Conceptual Apparatus

In reading the touchscreen as a skin, I suggest that the screen/skin interface has reshaped music listening because of how we relate to our skin and the skin of others. I deploy the conceptual apparatus of somatechnics that enables me to explore an understanding of skin (human or other) as a technology produced through bio-social-psychic structures. The body itself might be conceptualized as a technology, and as a technology, the body interacts with other technologies, both machinic and organic. For Dahl and Sundén, somatechnics is an intervention in critical inquiry that indicates "that technologies are not something that are added to bodies, but rather the means by which bodies and their politics are formed and transformed" (2013, 227). In the somatechnical model, listening is informed just as much by the technology of the ears, skin, and discursive apparatuses (such as language), as it is by machinic assemblages and digital networks. Somatechnics is valuable for this

[1] Animal skin does also work, i.e. a cat's paw. However, these devices are designed for human use, so I focus the 'humanness' of the skin-screen interface from here.

exploration because, as an investigative tool, it allows for the creative production of a variety of "open-ended" figurations in which to "imagine knowledge, bodies and subjectivities *otherwise* and in multiplicity" (227; original emphasis). For example, in this chapter, I imagine the bodies of both human and mobile media devices as deeply relational and indicative of the way bodies and their affects and/or emotions can be fundamentally changed by their contact with other bodies (human or otherwise). Thinking about mobile touchscreen technology using a somatechnical model therefore unlocks a variety of theoretical doors, through which the potentialities of music listening are coupled with the affective dimensions and emotional repercussions of *skin on skin relations*.

The field of somatechnics emerges from the legacy of feminist theories that sought to reimagine the body, in particular Haraway's cyborgian vision (Dahl and Sundén 2013, 227) in which hybridities of machine and organism serve as an ontology that might decolonize the body from so many political and social traumas. In particular, it cleaves from "the tradition of racist, male-dominant capitalism; the tradition of progress; the tradition of the appropriation of nature as resource for the production of culture; the tradition of reproduction of the self from the reflections of the other" (Haraway 1987, 2). Somatechnics advances and extends these ideas to build an understanding of technology as an "intimate part of what we have come to think of as our bodies" (Dahl and Sundén 2013, 227). In doing so, as Dahl and Sundén note, the core of somatechnics is about "border crossings" and "boundary confusions" (228).

It is the confusion and subversion of the screen/skin interface that concerns me here. I interrogate how these boundary confusions between screen-skin/human-skin can reconfigure emotional and affective dynamics in the listening experience, especially in regard to the way that affective dynamics of skin relations can mark expressions of love, tenderness, or even disgust and fear. Affect theory is deeply imbricated with models of somatechnical thought too. For Dahl and Sundén, affect constitutes the "somatechnical assemblages of images, media technologies, and bodies" (2013, 231). I take this approach as the launching point from which to explore and understand the way touchscreens of our mobile music media have changed ways of feeling in the digital listening experience.

Music Listening at the Border of the Human/Computer Interface

The touchscreen-skin mimics our own skin. The touchscreen is sensitive to stimuli, it reveals internal dynamisms, and it is smooth to the touch. This produces a kind of mirroring process by which the skin of the touchscreen, its 'face' if you will, becomes anthropomorphized in similar terms as the human body and human face. In online discussions, for example, we see concrete instances in which users personify their touchscreen devices in ways that suggest the skin-on-skin relation is critical in the affective experience of mobile media. For example, one Reddit user OtisDElevator explains that:

> Way back when I was at university, I was talking about a programming problem with another student. He kept touching my screen and leaving smudges. The third time he touched it, and after firm two warnings, I simply wiped my thumbs down each side of my nose and pressed the grease loaded thumbs on to his spectacles, saying, 'There, now how do you like it?' (2015)

For OtisDElevator, the memory of this 'screen-skin violation' still resonates years after the experience. This emotional narrative betrays a deeper condition of the human-device interface in which the interface has been absorbed into the matrix of the human skin-psyche. For OtisDElevator, the touchscreen skin is his/her skin, which is why after 'two firm warnings,' OtisDElevator feels justified to press 'grease-loaded thumbs' onto the glasses (another technology that extends the bodily surface) of the other student. These surfaces work as our own surfaces; they are personal and produce affects as would the organic skin. In response, another Redditor remarks, "I would never touch the screen of someone else's computer with anything... it just violates some moral code" (tylerni7 2010). Screens have been absorbed into the integrity of the human body dynamic. They require guardianship as with our very own organic body would and the border between what is our skin and the skin of the machine is no longer clearly demarcated. This resonates with the work of William J. Mitchell in work *Me++: The Cyborg Self and The Networked City* (2003) who describes the skin-covered body as only one aspect of the entirety of bodily experience:

> My natural skin is just layer zero of a nested boundary structure. When I shave, I coat my face with lather. When I'm nearly naked in the open air, I wear—at the very least—a second skin of SPF 15 sunblock. My clothing is a layer of soft architecture, shrink-wrapped around the contours of my body. (Mitchell 2003, 7)

In the experience of the externalized skins, the human body is extended and expanded in new somatechnical figurations. The technology of the skin—its ability to fold, produce sweat, stretch and absorb substances—is mirrored in the technology of mobile devices that are sensitive to touch, that can be broken or scratched, and that can be violated by the fingers of another. This is why using touchscreen devices is not just a one-dimensional listening experience, rather, it is an activity that calls forth all the sensitivity of skin-on-skin relations and attendant bodily integrities.

The function of skin as a surface (rather than organ, for example) is reversible too (Shildrick 2003, 165); one cannot touch something or someone without also experiencing touch themselves. In relation to touchscreen technology, when a user touches the screen to enact a function, the user is also being touched *by* the screen. This produces its own effects in which the user becomes intertwined with listening experience through the subtle sensualities of contact and caress. For psychoanalyst Didier Anzieu in *The Skin Ego* (1989), the surface of the body is the very basis of all experience, the "mental image of which the Ego of the child makes use during the early phases of its development to represent itself as an Ego containing psychic contents, on the basis of its experience of the surface of the body" (1989, 40). Our skins produce our corpo-*reality* and, thus, the skin of the touchscreen produces its own realities, or rather, we project onto it the limits and codes by which we understand embodiment to take place. The touchscreen makes music listening into a sensual, skin-on-skin phenomenon in a way that was not evident in traditional music listening. Anzieu's "skin-ego" also calls upon Merleau-Ponty's notion of "intercorporeality" which emphasizes "that the experience of being embodied is never a private affair, but is always already mediated by our continual interactions with other human and non-human bodies" (Weiss 1999, 5). As Dahl and Sundén explain, Merleau-Ponty's work "attends to the multiplicity of sense perceptions: bodies can be touched as well as seen" (2013, 5). In the instance of touchscreen mobile devices, music becomes embodied—it can be touched, seen, and be moved around as an extension of one's own body. Where the core body goes, so the touchscreen follows. It's another layer of the body which produces psychic investments in the music itself because that music becomes, also, another layer of the self.

Finally, the music itself is a thread that sews together the flesh of the human with the internal workings (or organs) of the mobile music device. The earphones work as a vein, 'pumping' the music through the body of the device to the body of the subject. I use the phrase 'pumping' the music deliberately here because it is a common phrase in street vernacular that also resonates with the language of the hydraulic model of the body as a system of pressures and conduits through which the blood circulates oxygen. The earphone cord becomes an externalised vein that fuses the human/device together as one in a somatechnical fusion. Mitchell's elaboration of the wearable self provides further insight into the way technologies are integrated into the self. In Mitchell's terms, the individual's body is "only the core" of the entire self:

> I consist of a biological core surrounded by extended, constructed systems of boundaries and networks. These boundary and network structures are topological and functional duals of each other. The boundaries define a space of containers and places (the traditional domain of architecture), while the networks establish a space of links and flows. Walls, fences, and skins divide; paths, pipes, and wires connect. (2003, 7)

The body is 'only the core,' and the MP3 player or tablet technology is linked to the self in a way that becomes a part of self and integrated into the somatechnics of the human body. Listening, in this case, is an integrated and holistic phenomenon in ways that traditional modes were not. The listening experience has reached a new somatechnical figuration here, in that bodily affect and discursive constructs of the body—what it means to be a connected self/device—generates an exciting new way to understand how music breaches outer surfaces.

We may think of our own body as a somatechnical figuration of a listening device in its entirety. For example, bone-conducting headphones now mean that the bone of the skull holds and channels the music through the head itself: "When hearing through bone conduction, sounds from the air or those presented through vibratory devices strike the bones of the skull bypassing the ear canal and stimulate the fluids of the cochlea" (Henry et al. 2009, 1). The music is literally infused inside the *bodily fluids*. The body becomes the device; our organs serving as the hardware, and our mind the software that runs the scripts. For Mitchell, in a post-biological future, we may think of ourselves as software. Popular discussions suggest that perhaps we are already there, in that our mobile listening devices have become compatible doubles of the self, that store our music, memories and structure ways of feeling. The self and the device are compatible. As Mitchell explains, this does not have to mean disembodiment, or the devaluation of the material self. Rather:

> It is a more complex, spatially distributed, fluid, hybrid form of embodiment enacted with new hardware— one in which silicon, copper, and magnetic subsystems play a vastly increased role, while carbon-based subsystems play a diminished and no longer so privileged one. Mortality reappears as a server crash. (There are some work-arounds, perhaps; you could implement reincarnation as restoration from backup, and transmigration of the soul as a hardware replacement strategy.) (2003, 168)

The body (*soma*) and the technology (*techne*) have become fused by way of the listening technology interpenetrating the body's own systems. The body has open channels allowing a constant flow in and out; it is by no means a closed system. The way music enters through the body is penetrative and intimate in psycho-sexual parameters suggesting stimulation of bodily fluids and corporeal resonances, vibrating our very bones in order to make the music manifest in new ways.

Sartre and the Erotics of Touch as Possession

The question that undergirds this investigation is why it is that touch should produce such intense and intimate relationships between human and device. Here I draw from Jean-Paul Sartre's phenomenology explored in his work *Being and Nothingness* ([1943] 1992) in order to illustrate how the touchscreen has intensified listening experience as a result of the acts of stroking and caressing, usually reserved for those things onto which we project desire. Firstly, Sartre points out that, if to stroke a surface was merely for the sole purpose of interacting with only the surface, there would be no relation between the surface and the stroke, no implication of desire or wanting (390). Portable media players fit into this model of desire in that the touchscreen device, particularly the iPod, is an icon of consumption obsession (see Kahney 2005). As Sartre writes, to stroke or caress an object of desire is not merely a superficial activity but a "shaping"

(1992, 390). He writes: “In caressing the Other I cause [his/her] flesh to be born beneath my caress, under my fingers” (390). Touch brings the body of the Other into the self and, by doing so, brings that body into knowing. It traps the Other into the material world of the self through the “game” of touch (Deutscher 2003, 144). In his work, Sartre is discussing the desire for the female form, however, this model can be applied in an understanding of the interaction between touchscreen device and listener because similar processes of desire take place in commodification culture. To take Sartre’s approach, the production of music listening pleasure is ultimately tied to the appropriation of the Object through touch. As Penelope Deutscher explains: “touch occurs as part of a subject’s making itself body. For Sartre, the touch of a skin against an object or another occurs as part of a constant project to appropriate the world” (2003, 146). Touch, catalyzed from the desire to attain a form of pleasure (which can be read here as either sexual encounter or musical experience) implies a much deeper dimension of affective phenomena than just the interaction of boundaries and surfaces.

To follow from this line of inquiry, the listener too can be said to appropriate all the potentialities and sensualities that the device contains, as would a lover. The actions and activities of finding the music (stroking the screen), gently pushing its surface (to select songs) and so forth, produce the pleasurable relationship with the touchscreen device. Even the language of the mechanics to turn the device ‘on’ or ‘off’ is suggestive of the sensual act in progress. As Sartre goes on to say:

> My shirt rubs against my skin, and I feel it. What is ordinarily for me an object most remote becomes the immediately sensible; the warmth of air, the breath of the wind, the rays of sunshine, etc.; all are present to me in a certain way, as posited upon me without distance and revealing my flesh by means of their flesh. (Sartre 1992, 392)

What Sartre is saying here reminds me of the way the touchscreen device provokes similar effects of being-in-the-world because it has both immediacy, ‘without distance,’ and also reveals the listener’s flesh to be in the action of appropriating pleasure from the surface of its skin.

This experience is not like that of the radio, where music is filtered through the air, nor is it like the experience of live music which produce completely different effects to do with collectivities. This experience is specific to the function and erotics of touch and the immediacy that the touchscreen device can sustain. Therefore, the question as to why touch, in particular the type of touch-enabled in mobile touchscreen devices, produces such intense relationships between human and device, is answered by Sartre's understanding of the way stroking, caressing and other modes of touch can engage the listener in sensorial and sensual encounters with erotics of consumption. Deustcher goes on to explain that in Sartre's model, to desire (as the listener desires the touchscreen) "is first to be reduced to one's own body" (2003, 143). Deustcher quotes Sartre in saying that to desire and to caress as a function of that desire enables one to feel one's own skin, one's own muscles and one's own breath: "I feel them not in order to transcend them *toward* something as in emotion or appetite ... but as a *passion* by which I am engaged in the world and in danger in the world ... The being which desires is consciousness *making itself body*" (as cited in Deutscher 2003, 143). The touchscreen device brings the listener *into the music* and by doing so, into the erotics of listening. In Sartrian terms, erotics is present in its facticity, "the contact of two skins" (1992, 56). I suggest that, in the digital age, two skins may involve very different versions of the human epidermis. Touching the screen-skin, manipulating the controls and appropriating the music as pleasure is not a mere superficial encounter. Rather, it is an intense interaction between two concrete forms that produce the world within that interaction.

Haptics: The 'Touchy-Feely' World of Mobile Music

Sartre's erotics of touch provides a way of understanding how the tactile bond between user and device is cultivated in relation to the praxis of desire. However, this still does not address the question of why that tactual bond should be so in demand as popular music culture *appears* to be moving towards more immaterial modes of listening. In Chap. 2 I emphasized the importance of the tangible world of music listening as it has been cultivated and naturalized through twentieth century practices. Here, I argue that, while materiality has become certainly less prevalent in some music listening practices (streaming, for example), the need for the tactual bond does not disappear, rather, it is translated into the model of touch as it is specific to that of haptic technologies. As Ahmed points out in *The Cultural Politics of Emotion*, "the lived experience of being-itself depends on the intensification of the skin surface" (2004, 104). This dependency has not vanished completely, instead, it has become refocused. However,

I do not suggest that haptic interfaces, such as the touchscreen, provide the same types of skin-on-surface contact that characterized twentieth century relations with tangibility. While twentieth century models of tactual bonds were about physicality, haptics is about *closeness.*

Closeness is critical to the practice of music listening because listeners are always in the process of bringing music 'close.' Music is not here. Music is 'otherwise-here'—in our ears, our minds and our bodies. Music is not here, but it is *almost* here, so in this sense, music is virtual. The techniques of touch in haptic-listening demonstrate the reach for bringing the virtual into the here, or giving the virtual a sense of 'here-ness'. For example, love is a virtual experience in that it cannot be said to be in a specific place. Therefore, in order to make love 'concrete,' to bring love closer to a material experience, we bring love into the physical by manifesting the virtual into something we can touch—such as a wedding ring, a kiss on the mouth of the loved one, the ideographical heart carved into a tree. In terms of music, better audio quality brings the music 'close' by enabling a deeper appreciation of the attributes of that music. Live concerts bring the music close by adding extra sensorial dimensions to the listening experience, such as interaction with artists and other fans. Listening to music on an LP brings the music close because the listener can connect physically with the weight of the vinyl, the artwork on the cover and even the smell of the sheath.

Touchscreen devices do not provide any of those things. What they do provide, however, is all of one's music available at the touch of the screen. I argue here that this capability is not just about convenience, rather, it is also about closeness. The listener becomes immersed in their collection, the virtual collection is pulled into existence at the manipulation of the touchscreen, it is brought closer through gentle nudges and grazes of the user's fingers. The music is all there, all the time, simultaneously. As Mark Paterson explains, the differences in computer interfaces produce radically difference degrees of closeness, "Whereas the keyboard is a passive mechanical channel between the computer and user, haptics enables a more active exploration and allows the user not just to *see* three-dimensional shapes represented on the screen, but also to *feel* them and interact with them" (2007, 374; original emphasis). The mobile device listener is embedded within the virtual matrix of their music collection that exists with them and that listener can move through that virtual matrix as if they are native to it. For Paterson, haptic interfaces such as these, "are a set of augmentations that begin to play with an emerging multisensory realm, one that talks of the engendering and engineering of 'immersion,' of 'presence,' of 'aura' through the addition of touch" (2007, 374). Paterson goes on to explain that it is the sense of copresence that "fosters feelings of nearness and intimacy" (374). Similarly, the mobile device user is copresent with all of their music, which enriches the sense of intimacy with the music experience in ways that were not possible in more traditional listening practices. Having one's entire collection 'at one's fingertips,' in the language of mobile device marketing, is therefore about much more than a convenient way to access and transport large amounts of music. It is also about the interweaving of one's experience with the virtual world of music, bringing that music closer into the experience of the being.

If the outer layer of the screen works as a kind of epidermis then we can read the inner mechanisms of the device as internal organs, suggesting an interiority to the device that produces even further sensualities from which to read affectivity in relation to mobile media. Specifically, these sensualities are produced via the operation of absence and presence—the mobile device as prosthesis. As I illustrate above, the touchscreen device is implicated in intimate relations with the human body; it is connected to the body and mirrored by the skin. Therefore, I read the touchscreen device as a prostheticised cyborgian figuration. Its face is sensitive to the touch, and by touching it in the right ways with the right pressures, music is released, and pleasure is produced. Sensual and sexual relationships between the human and the cyborg have long been part of the cultural psyche. For example, there is a recurring theme in popular culture in which sexual relationships between human and cyborg are imagined and explored (see films such as *Blade Runner*, *Cherry 2000*, and the film *A.I.* which also features cyborg prostitution). These films explore modern anxieties, as well as the *frisson*, that is produced in the development of relations between techno-machinic devices and individual bodily boundaries. As Sharalyn Orbaugh notes, "explorations of cyborgianism are not about the future but are the expression of current ideas about the human body" (2002, 436). For Jennifer Gonzalez too, "The image of the cyborg body functions as a site of condensation and displacement. It contains on its surface and in its fundamental structure the multiple fears and desires of a culture in the process of transformation" (1995, 267). The touchscreen device, as cyborgian figuration, invokes confusion and fusion between the subject and machine. This produces new affective attachments seated in both fleshy corporeality and discursive meanings about how one defines the borders of affection and its role in music listening. Listening to music using a mobile touchscreen device involves a far more complex emotional and visceral schema that, as suggested here, call forth feelings of bodily territoriality and psychic borders that maintain the music device, and the music contained within, as an *extension* of the human sensual form.

Music Players as Sex Toys: I'll Never Listen to the Black Eyed Peas in the Same Way Again

Over the past ten years, the adult market has released musically-powered vibrating sex toys, based on the iPod brand, which can be manipulated via a touchscreen (Gluckstern 2010). This enables a direct sexual release and even orgasm using the mobile music experience:

> the Ohmibod is a music-powered vibrator that translates tunes from any MP3 player (actually any audio source so iRiver, Creative, Zune etc will all work too) into vibrations. Offered as an 'acsexory' aimed squarely at the world's largest selling personal entertainment device (the Apple iPod). (Hanlon 2006)[2]

The music itself drives the force, intensity, rate and speed of the vibration cultivating an acutely intimate relationship with the music listening experience. This device implies a somatechnical coupling between human and machine on multiple levels, and in ways that radically redefine the affective dimension of music listening. What I mean is that the technologies of the cis-female genital organs work together with the technology of the device in order to produce feelings and sensations that were not previously possible with music listening (the device is marketed toward cis-females, but males could deploy the device for their own pleasure). The somatechnics of this relationship suggests a radical redefinition of how music shapes the body and in what ways bodies can be manipulated by the mobile music device. As one reviewer of the product writes, "I'll never listen to the Black Eyed Peas in the same way again" (Hanlon 2006). The ways in which we experience intimacy, pleasure and connection in the music listening activity has been ultimately and radically reshaped in ways that traditional mediations were not.

[2] There are also two other similar products on the market to date, iBuzz and iGasm.

Musically-powered sex toys therefore imply a new space for imagining female sexuality in terms that invite the coupling of music and genitalia, personalizing and perhaps even empowering music listening in gendered terms. However, the relationship between female sexuality and technology is as complex as it is controversial. As Donna Drucker covers in her work on *The Machines of Sex Research* (2014), in the late Victorian era machines were used to provide women "relief from 'hysteria'" and surgeons "use certain tools to shape or to 'fix' genitals into idealized contours and sizes" (1). Female sexuality has been utterly colonized by the medical profession and continues to be regulated and controlled by the machines and technologies of scientific rationalism. However, there have been and continue to be forces that pull in other directions, which serve to redistribute the balance of biopolitics in favour of women's right to bodily integrity. For example, Drucker goes on to explain that the emergence of second-wave feminism, gay and lesbian rights, and civil rights catalysed a shift in sexual biopolitics "from punitive measures to investigations of basic physiology and numerous other topics" (2). She writes that:

> Radical feminism galvanised the American feminist movement in the late 1960s and early 1970s, cementing key concepts such as 'the myth of the vaginal orgasm' and 'the personal is political' into popular discourse and consciousness [Faludi 2013]. Women worldwide made numerous political, economic, and cultural gains from the late 1960s through the mid-1980s, including the passage of Title IX in the U.S. requiring gender equality in college sports, contraception becoming free in the United Kingdom. (Drucker 2014, 2)

I cover this history briefly in order to contextualize the biopolitical implications of the musically-powered vibrator, insofar as it relates to the somatechnics of women's experiences of music as they have shifted from traditional listening to new digital paradigms. What this context provides is a far more heterogeneous narrative of affect and encounter; the musically-powered vibrator proposes an empowerment but one that has emerged from the histories of scientific-patriarchal paradigms. In *The Technology of Orgasm*, Rachel Maines explores the "medicalizing of female orgasm in Western culture" that is deployed to protect dominant and patriarchal "illusions of coitus" (1999, 121). For instance, the musically-powered vibrator is shaped as what Maines would describe as the "reassuringly phallus-shaped vibrating dildo" which can be understood as a substitute for the penis. This suggests that the device serves not only the technologies of female genitalia but also the paradigm of male sexuality that requires females to continue to desire the phallus, in the case that vibrators should render the male subject unnecessary.

In a somatechnical model, the music-powered-vibrator therefore implies a complex negotiation between the technologizing of the female orgasm and the appropriation of contemporary models of digitization for female empowerment. For instance, the musically-powered vibrator could be read as an exploitation of female sexuality to sell new technologies or it could be read as a new model for women to integrate music into their sexualities more deeply and powerfully. I do not propose here a resolution that comfortably ties together the political ramifications of such a debate. However, what is evident is that the development of touchscreen technologies in mobile music media are producing new strategies of listening that work in conjunction with human sexuality, suggesting the touchscreen as a type of cyborgian sexual figuration.

Grief and Loss in the World of Mobile Touchscreen Technology

As I have been suggesting, the touchscreen skin is, in a way, alive, particularly in the minds of users. It responds to the living-ness of the human skin's electrical conduction and responds to the commands it *feels* through changes in its own skin's electrostatic field. As a result, it works as a kind of cyborgian prosthetic which produces new forms of pleasure and intimacy in the music listening experience, but also new possibilities for absence and loss. In this section I explore the phenomenon of grief in regard to mobile media and inquire as to what this phenomenon can tell us about the specificity of listening experience in digital contexts. I suggest that the relationship between the touchscreen-human interface, which I have explored above, results in the treatment the touchscreen device as a thing that can experience death, just as the skin itself might experience death through trauma, old age, or conditions like rash or gangrene. The device is so attached to the human body that it ingratiates itself as a kind of prosthetic instrument, one that is vulnerable to loss, theft, and the precariousness of the outside world. It is symbiotic with the human, like a fetus. It requires charge in order to sustain its life and the human requires it to be living in order for it to provide listening pleasure. When that relationship ends, it is a kind of death. As a result, popular music communities produce strategies of mourning in order to cope with the loss of these personal devices. For example, in the Reddit community, language is used that indicates the collective grieving in regards the discontinuation of the iPod Classic: "She's finally gone [...] Good Bye iPod classic" and "RIP iPod Classic" (8lueBuddha 2014). Even though the iPod Classic did not have a touchscreen display, it was the original device that led to the entire suite of touchscreen technologies for portable media players that followed (Hiltzik 2014) and had a 'click wheel' that could be stroked and manipulated. Users personify and grieve the obsolescence or loss of all types of technologies, from cars to CD players to tennis rackets. However, the portable media player is grieved and personified in specific ways that differ from other technologies because it is the first technology of its kind to have the capacity to hold such a vast amount of music, to be robust enough to travel anywhere without skipping or damaging the internal CD or cassette, and generally to feature as a constant companion to the modern music listener as in a prosthesis which can be mourned if lost.

To frame the mobile device as a prosthesis, the term 'prosthesis' is used here as "a metaphor signaling some kind of mediation between an artificial device and the supposedly natural body" (Shildrick 2009, 133). As Katherine Ott points out, it is a "literal interface between flesh and machine" (2002 as cited in Shildrick 2009, 133). Mobile devices, particularly those devices that can receive commands via voice activation or alternative controls attached to the earphones, work as prostheses which enhance the listening experience by enabling the freedom of music to be heard and controlled anywhere the user desires. Bull attends to the high regard that users have for this kind of prosthesis, particularly office workers who bear the everyday commute by using these functions (2005, 348). This is reminiscent of how Shildrick writes about prosthesis. She says, "Historically, such technologies [prostheses] have usually grounded some utilitarian compensation for a perceived bodily lack, but the emphasis now is firmly on enhancement and supplement" (2015). In this understanding, the mobile device is not a technology that needs to bridge a lack or repair damage, but it is a technology that can enhance and supplement any human body and bring relief from the demands of the outside world. Shildrick writes, in the current context, "issues of bodies, boundaries, and technologies increasingly challenge not only the normative performance of the human subject, but also the very understanding of what counts as human" (2015, 13). Rather than considering prosthesis as a management tool to repair the 'non-acceptable' human body, prosthesis here is read as a *concorporeality* (Shildrick 2009) between the "organic and inorganic, the assembly and disassembly of surprising connections, the capacity to innovate" (Shildrick 2009, 133). Prosthesis is therefore about potentialities.

With the above statements in mind then, the mobile device as prosthesis relates to practices of mourning because the death of the device points to the end of a part of the body that one had invested with all the potentialities of listening pleasure, which itself constitutes a kind of living-ness. When read in this way, it is no wonder that individuals grieve the loss of these devices as they would with any integral bodily operation that offers life-enhancing features. In a discussion about mobile music players online, one Reddit user expresses a kind of grief by explaining that his/her iPod "recently died after about seven years of faithful service" (yanchovilla 2014). This device was ingratiated into the life of this user. The impact of grief on the listening experience in digital contexts is therefore interesting as it creates a discourse around loss and technology. Mostly, this means discussions and testimonials of one's device to others in the music community. Another Redditor dfloyd13 begins a group discussion with a thread title borrowed from a *Wired Magazine* article "'Looking at someone's iPod was like looking into their soul' On death and iPods: A requiem" (dfloyd13 2014). The thread received 287 responses, many of which were commiserating the loss of personal devices, such as k_thrace who responds with: "Somebody stole my soul out of my car then" (see dfloyd13 2014). Also in response, Redditor justicecupcakes, says: "I can't really imagine ever going without mine. It's a bit pathetic, but it's a major part of my life" (see dfloyd13 2014). This Redditor demonstrates the extent to which the mobile device (in this case the iPod) can support the emotional microcosm of the user, so much so that one 'cannot imagine to live without it'. In Bull's words, the iPod works as a "'framing' device, enabling a distinctive mode of auditory embodiment—governing the way in which iPod users engage and orientate themselves to the world and to themselves" (2008, 22). Bull's vision resonates with the notion that the iPod device (or any mobile device with similar capabilities) becomes an extra-sensory prosthetic instrument in which experience is redefined. The device recalibrates the user's interaction with the external world using cyborgian strategies in which "the iPod user is an 'orchestrating' self who tones down stimuli from one sensory field [whilst] amplifying information coming through another channel" (Geurts 2002, as cited in Bull 2008, 22). The outside world—that is, the world outside of the user's listening experience—can be ciphered and then modified through the parameters of the technological mediation at work.

This echoes Kathryn Geurts' assertion that the way an individual experiences sensorium is by no means a natural phenomenon untouched by cultural scripts and codes. Rather, Geurts explains that sensory and somatic practices only become embodied through the "ushering into" that culture's sensorial framework (2002, 232). The body must learn how to interpret the cultural sensorium in order to make sense of signals surrounding it. Using a portable media player requires a refashioning of this interpreting schema and, in this way, the body's somatechnic—its sensorial faculties and cognitive abilities—works symbiotically with the prosthetic of the portable media player in order to cultivate the peak listening experience in a mobile setting. In this sense, the human body inheres the cyborgian technology. As a result, I suggest that digitisation enables the mobile media device unprecedented capacities, that offer not only convenience and fluidity but also a new way of being that reconfigures human/cyborg relations in new and intimate ways.

Conclusion

I have filtered the touchscreen through a variety of conditions in this chapter. Its skin has been exposed to the molestation of sticky fingers and jabbing nails. Its body has been appropriated through Sartre's erotics of touch and caress. It has been tethered to the somatechnics of female genitalia and sexual gratifications. Users have mourned the passing of their devices, and the devices of others. Chronic use has suggested the device as a prosthesis that transforms somatechnics of sensorium and the world outside the 'iPod bubble'. Users have absorbed these devices so far into the fleshy folds of the human skin that the borders between prosthesis and subject have given way to a cyborgian figuration in which new models of feeling can be explored and imagined. Specifically, these are ways of feeling that extend bodily boundaries, as well as integrate music listening into the very fluids of the body. As Claudia Castañeda writes, "Human nature as it is investigated, generated and lived, is said to be undergoing a transformation that explicitly breaches the human/non-human divide" (2003, 223). The mobile device sits at the precipice of this divide.

The overarching theme of this chapter, however, is the skin and the way that the touchscreen device specifically produces new somatechnical figurations based on skin-on-skin relations. The skin demarcates bodily borders based on the ontology of surface and containers, meaning the skin is imagined as the border between the self and everything else which is not the self. The psyche generates its corporeal boundaries based on the edges and borders of that skin, or what Anzieu named "the skin-ego" (1989). The skin is far from a superficial utility. It is a locus of bio-social-psychic processes, and produces a *frisson*.

In this chapter, I have also constructed the touchscreen device as an inorganic Other, onto which the user can project a variety of feelings and notions about pleasure and possession. The fact that the touchscreen device, as cyborgian construction, does not have consciousness or agency is not relevant here. Social constructions of love, desire, erotics and pleasure often rely on the notion of two autonomous subjects. However, to scratch this surface even slightly is to reveal a wealth of relationships between human and non-human or non-conscious figurations which produce just as intense and profound ways of feeling as any other. For Castañeda, "To ask whether robots have skin is to ask about 'our' posthuman nature and its embodiment as it is being re-imagined in technoscientific domains" (2003, 223). This being the case, I reimagine the touchscreen-skin relationship as a new way to experience listening as a techno-concorporeal and prostheticised figuration at the interface of human-computer relations. By doing so, I interrogate the theoretical and material limits that demarcate human and machine in the scope of music listening. The human is a technology; albeit a fleshy organic technology that operates within the natural physical laws of growth and decay. However, the boundaries of the human body are not set. They are fluid, adaptive, elastic, mutable and most of all, interconnective. The human body can be hooked up to a machine to monitor heart rate, the body can be hooked up to another machine to support its very life force. In these instances, the human body is only a node in a series of technics which extend experience and interconnect with organic and inorganic processes.

Reading the mobile touchscreen player as a somatechnical figuration therefore suggests that the listening experience is developing along with the technologies that mediate music to the body in ways that continue to challenge our understanding of bodily borders and in ways that redefine what it means to *feel* the music. Therefore, the touchscreen-skin is a critical site of affective relations that dramatically reshape what it means to listen to music in a mobile setting; a private and intimate encounter between the user and their counterpart.

References

8lueBuddha. 2014. She's Finally Gone. *Reddit.com*. https://www.reddit.com/r/apple/comments/2fxm8o/shes_finally_gone_good_bye_ipod_classic/

Ahmed, Sara. 2004. *The Cultural Politics of Emotion*. Edinburgh, Scotland: Edinburgh University Press.

Ahmed, Sara, and Jackie Stacey. 2003. *Thinking Through the Skin*. Hoboken, NJ: Taylor & Francis.

Åman, Pirkka, and Lassi Liikkanen. 2013. Painting the City with Music: Context Aware Mobile Services for Urban Environment. *Continuum* 27 (4): 542–557.

Anzieu, Didier. 1989. *The Skin Ego*. New Haven, CT: Yale University Press.

Beer, David. 2010. Mobile Music, Coded Objects and Everyday Spaces. *Mobilities* 5 (4): 469–484.

Bull, Michael. 2000. *Sounding Out the City: Personal Stereos and the Management of Everyday Life*. Oxford: Berg.

———. 2005. 'No Dead Air!' The iPod and the Culture of Mobile Listening. *Leisure Studies* 24 (4): 343–355.

———. 2008. *Sound Moves: iPod Culture and Urban Experience*. New York: Routledge.

Castañeda, Claudia. 2003. Robotic Skin: The Future of Touch? In *Thinking Through the Skin*, ed. Sara Ahmed and Jackie Stacey, 223–236. Hoboken, NJ: Taylor & Francis.

Cataldi, Sue L. 1993. *Emotion, Depth, and Flesh: A Study of Sensitive Space, Reflections on Merleau-Ponty's Philosophy of Embodiment*. Albany, NY: State University of New York Press.

Dahl, Ulrika, and Jenny Sundén. 2013. Guest Editors Introduction: Somatechnical Figurations. *Somatechnics* 3 (2): 225–386.

Deutscher, Penelope. 2003. Three Touches to the Skin and One Look: Sartre and Beauvoir on Desire and Embodiment. In *Thinking Through the Skin*, ed. Sara Ahmed and Jackie Stacey, 143–159. Hoboken, NJ: Taylor & Francis.

dfloyd13. 2014. 'Looking at Someone's iPod Was Like Looking into Their Soul' On Death and iPods: A Requiem. *Reddit.com*. https://www.reddit.com/r/gadgets/comments/2gy4xu/looking_at_someones_ipod_was_like_looking_into/

Drucker, Donna. 2014. *The Machines of Sex Research: Technology and the Politics of Identity, 1945–1985*. Dordrecht: Springer Netherlands.

Fanon, Frantz. [1952] 1967. *Black Skin, White Masks*. Translated by C. Lam Markmann. New York: Grove Press.

Geurts, Kathryn Linn. 2002. *Culture and the Senses: Bodily Ways of Knowing in an African Community*. Berkeley: University of California Press.

Gluckstern, Nicole. 2010. The Performant: Dies Arses Have It. *San Francisco Bay Guardian*, December 31. http://www.sfbg.com/pixel_vision/2010/10/06/performant-diesarses-haveit

Gonzalez, Jennifer. 1995. Envisioning Cyborg Bodies: Notes from Current Research. In *The Cyborg Handbook*, ed. C. Hables Gray, Steven Mentor, and Heidi Figueroa-Sarriera, 267–279. New York: Routledge.

Gram, Nina. 2009. Music Mobilizing Meaning within Place: An Examination of Different Central Characteristics of the Mobile Sound and Listening Space. Paper presented at *Motion and Emotion within Place: International Conference*, Aarhus, Denmark, October 7–8.

Hanlon, Mike. 2006. Ohmibod: The First iPod Acsexory. *Gizmag*, December 31. http://www.gizmag.com/go/6492/

Haraway, Donna. 1987. A Manifesto for Cyborgs: Science, Technology, and Socialist Feminism in the 1980s. *Australian Feminist Studies* 2 (4): 1–42.

Henry, Paula, Phuong Tran, and Tomasz Letowski. 2009. *Comparison of bone conduction Technologies Technical Report*. Aberdeen, MD: Army Research Lab Human Research and Engineering Directorate.

Hiltzik, Michael. 2014. RIP: The Apple iPod, 2001–2014. *LA Times*, December 31. http://www.latimes.com/business/hiltzik/la-fi-mh-rip-the-apple-ipod-20012014-20140911-column.html

Kahney, Leander. 2005. *The Cult of iPod*. San Francisco, CA: No Starch Press.

Kirk, David. 2012. Why Can't I Use Gloves with My Touchscreen Phone? *TechRecipes*, December 31, 2015. http://www.tech-recipes.com/rx/20564/why-cant-i-use-gloveswith-my-touchscreen-phone/

Maines, Rachel. 1999. *The Technology of Orgasm: 'Hysteria,' the Vibrator, and Women's Sexual Satisfaction*. Baltimore: Johns Hopkins University Press.

Mitchell, William J. 2003. *Me++: The Cyborg Self and the Networked City*. Cambridge, MA: MIT Press.

Orbaugh, Sharalyn. 2002. Sex and the Single Cyborg: Japanese Popular Culture Experiments in Subjectivity. *Science Fiction Studies* 29 (3): 436–452.

OtisDElevator. 2015. Comment. *Reddit.com*, December 31. https://www.reddit.com/r/reddit.com/comments/b20xa/do_you_hate_it_when_others_physically_touch_yours/c0kks81

Paterson, Mark. 2007. *The Senses of Touch: Haptics, Affects and Technologies*. Oxford: Berg.

Sartre, Jean Paul. [1943] 1992. *Being and Nothingness*. New York: Washington Square Press.

Shildrick, Margrit. 2003. 'You Are There, Like My Skin': Reconfiguring Relational Economies. In *Thinking Through the Skin*, ed. Sara Ahmed and Jackie Stacey, 160–174. Hoboken, NJ: Taylor & Francis.

———. 2009. *Dangerous Discourses of Disability, Subjectivity and Sexuality*. New York: Palgrave Macmillan.

———. 2015. Why Should Our Bodies End at the Skin?: Embodiment, Boundaries, and Somatechnics. *Hypatia* 30 (1): 13–29.

Sterne, Jonathan. 2006. MP3 as Cultural Artifact. *New Media and Society* 8 (5): 825–842. https://doi.org/10.1177/1461444806067737.

Sullivan, Nikki, and Samantha Murray, eds. 2009. *Somatechnics: Queering the Technologisation of Bodies*. Surrey: Ashgate.

Tyler, Imogen. 2001. Skin Tight: Celebrity, Pregnancy, and Subjectivity. In *Thinking Through the Skin*, ed. Sara Ahmed and Jackie Stacey, 69–83. Hoboken, NJ: Taylor & Francis.

tylerni7. 2010. Comment. *Reddit.com*, December 31, 2015. https://www.reddit.com/r/reddit.com/comments/b20xa/do_you_hate_it_when_others_physically_touch_yours/c0kkmg2

Weiss, Gail. 1999. *Body Images: Embodiment as Intercorporeality*. New York: Routledge.

Werner, Ann. 2015. Moving forward: A Feminist Analysis of Mobile Music Streaming. *Culture Unbound: Journal of Current Cultural Research* 7 (2): 197–213.

yanchovilla. 2014. She's Finally Gone. *Reddit.com*. https://www.reddit.com/r/apple/comments/2fxm8o/shes_finally_gone_good_bye_ipod_classic/cker0z2

CHAPTER 5

The 'Creative Listener:' Internet, Music, and the Computer-Bodymind Somatechnic

The human body interacts with machines in many ways. Many of these ways are obvious, but none are ever simple.
Rawdon Wilson (1995, 240)

It is the freely imaginative mind that is at the core of all vital music making and music listening.
Copland (1952, 7)

Introduction

When we listen to music, we do not *only* listen to music. We ideate, we let our mind wander, we imagine ourselves in new situations unfolding with the music, we daydream, we let the music affect our bodies and sexualities. We often couple music listening with other practices; we might utilize music to 'get our blood pumping' (as with exercise) or relax our autonomic systems (as with meditation or yoga). Some music is created quite deliberately as an accoutrement to love-making, and other music is quite deliberately produced as a soundtrack for psychedelic experiences, and the list goes on. The point, of course, is that the moment of music listening is always moulded quite deliberately with a variety of other techniques which help to reshape the experience of that music listening encounter. How this happens is a result of the listener's ability to listen *creatively* and it is the nature of our relationships with technologies that are the crux of how these encounters are built and manipulated.

L. Glitsos, *Somatechnics and Popular Music in Digital Contexts*, Pop Music, Culture and Identity,
https://doi.org/10.1007/978-3-030-18122-2_5

In this chapter I focus on the Internet as a technology of music listening as it is accessed exclusively through fixed-point personal computers. I am concerned with how the listener actively and creatively produces their own listening experience as it is mediated through this specific portal. This line of thought builds somewhat on the work of David Hargreaves, Dorothy Miell, and Raymond MacDonald's research in which they expound on the significance of "imagination and creativity in music listening" through which they view "perception as [the] creative construction of knowledge" (2011, 156). I take this further to argue that the coupling of the bodymind and the personal computer produces a *unified field of relations*, an artifact which we can examine through the lens of somatechnics. This unified field of relations, what I will refer to from now as the 'personal computer bodymind,' or for short the PCBM, realigns and reconstitutes listening pleasure in unexpected ways because it enables unique and creative listening practices contingent upon the functional nature of Internet-computing technologies as they are deployed in interactive Web 2.0 culture. The PCBM produces a new 'type' of listener: the creative listener.

New kinds of listening practices are always emerging; however, I will focus on the following practices specifically in order to track this argument: Creative 'geolistening;' community listening and the vaporwave phenomenon; and, finally, the role of algorithmic technologies in music discovery and playlisting. In relation to understanding these practices as somatechnologies, I suggest that creative listeners can mould and manipulate their listening practice (either consciously or unconsciously), taking their pleasure of music to unexpected places, which thus implies technicity. The creative listener is reflexive, open, exploratory, and engaged, and *brings forth* the terms of their listening pleasure by manipulating a variety of techniques, a process we might call the essence of *tekhné*. Heidegger calls to our attention that *tekhné* is:

> Not only handcraft manufacture, not only artistic and poetical bringing into appearance… it is a bringing forth, *poiesis*. *Physis* also, the arising of something from out of itself, is a bringing-forth, *poiesis* … *Techne* is the name not only for the activities and skills of the craftsman, but also for the arts of the mind and the fine arts. *Techne* belongs to bringing forth, to *poiesis*. (Heidegger [1954] 1977, 10)

The computer-bodymind coupling is a bringing forth of new creative listening practices, and ultimately, new affective encounters with music. The computer-bodymind coupling is also the exteriorization of the pursuit for musical pleasure, the constituting property of somatechnicity. However, the nature and expectation of these encounters are always predicated on the context of (technical and technologizing) forces which manufacture what kinds of pleasures are possible and valuable in any given era. For instance, while the pleasure of music listening may always have been social in some aspect, the nature, characteristics and quality of that sociality continues to adapt and change through the emergence of Web 2.0 software which, among many other things, fosters collaborative engagement over long distances. Users can also share entire personal histories through playlists on Spotify, they can recreate memories and validate relationships by sharing YouTube music videos on social media platforms. It will be these kinds of emergent strategies that I will examine in the following chapter in order to synthesize notions somatechnics with the pursuit of creative music listening practices in the digital context.

I want to emphasize, too, that using the Internet as a technology of music listening, *particularly* as it is accessed exclusively through fixed-point personal computers shapes, conditions and qualifies all dimensions of that music listening encounter. This is otherwise to state, in Marshall McLuhan's famous words, that "the medium is the message" (McLuhan and Fiore 1967). However, in the examples of the personal computer, how might we understand the medium as 'being' the message? First, we must recognize that in popular or common understandings, the relationship between the individual and the computer is generally regarded through the traditional 'human/tool' binary, in which the computer operates as merely an instrument of the human will, that is, as simply a medium which transmits the data indicated by the user in a neutral and direct fashion. This is erroneous, for computers do things we do not want them to do all the time. Computers, when turned on, are continuously running scripts and programs of which the user is unaware and in many ways, the individual is at the mercy of the computer (anyone who has experienced the frustration of an endless and unwanted Windows update will immediately understand this point, I'm sure). David Wills reminds us that, "at a moment in which the human appears to be moving inexorably forward towards a biotechnological future, it is strategically important to recognize—to be cognizant in return of—the fact of a relation between *bios* and *tekhnē* so complex and so historic that any presumption of the priority of one over the other can be sustained only by man's of an appeal to a metaphysics of creation" (Wills 2008, 5). Wills points to the nature of the human-computer coupling as only the most recent marker in a very long history of humanity's dependence on techné for being in the world and bringing forth any material conditions of that world. From the invention of the hammering device, to the most extant computing technologies, humanity is contingent on techné and on the technologization the human body. The pursuit of music listening with new and sophisticated emotional strategies is a product of the nature of this *poiesis*, or this exteriorization of listening pleasure—this bringing forth—which is ultimately tied to the medium itself.

In another sense, and as recent hacking culture has demonstrated, our personal computers are very much susceptible to outside forces and susceptible to threats imposed from external agents, very much like the vulnerable fleshy body itself. I am reminded here of the nude-photo hacking scandal in which personal photos of several female celebrities were exposed. This kind of incident pitches the personal computer as having porous boundaries, which are susceptible to the outside world (Durkin 2018). In this instance, the boundaries of the machine mimic and reflect the boundaries of the self, in that the personal computer holds the body (albeit the *image* of the body, which is the contestable artifact in this situation) in a manner which suggests the same vulnerability as the flesh. Despite this obvious vulnerability, the traditional 'human-tool' binary, in which the user enacts control over that tool, has been naturalized through the presupposition of human superiority and the 'invisibility' of technology, particularly technology that works without causing disturbance. In the words of Robert Rawdon Wilson, "machines are so omnipresent in the western technological environment that they are also invisible … while they are still functioning, the extended systems that make machines possible are largely out of sight, too complex to be seen easily" (1995, 241). To make visible these complex systems is to see that the human-computer interface is coadjuvant, in that one increases and augments the effects of the other.

This is not to imply agency within the computing system, but to indicate that the coupling of the computer/human is far messier, disordered and non-hierarchical than is usually allowed by conventional 'everyday' understandings. Though the computer is constantly running many processes to which the human is oblivious, the human, too, is running complex processes, including emotional, discursive, sensorial ones, on multiple levels, that are irrelevant to the processes of the computer. Both human and computer sustain simultaneous and deeply rooted pre-established scripts while also both maintaining open lines of communication to and from the other. Therefore, how we undertake processes of creative thinking in relation to listening practice is shaped and reshaped by this figuration as it is mediated through the computer-human interface. For instance, we might note the simple yet extreme changes in artistic expression as contingent upon technology in any given context. Nick Prior unpacks this relationship with a prime example of the productive practices emerging from the uptake of turntable technology across various sites of use. He writes:

> We only have to look at the history of the turntable to see how an object meant for reproduction (playing records) was transformed into an instrument of production through the localized, tactile appropriations of Jamaican and black American hip-hop DJs in the 1970s. In fact, the history of popular music is littered with examples of users redefining technology's functions and meanings, sometimes inadvertently, in the flow of everyday actions. The techniques of distortion, feedback, gated reverb, beatboxing and scratching can all be sighted through the prism of user contingencies, while the distinctive bass line 'squelch' of acid house in the mid 1980s is a case of what elsewhere I have termed the 'extemporising bodies' of musicians in Detroit and Chicago. (Prior 2018, 8)

While this example points largely to the ways in which technology and creativity co-articulate in the processes of music production, in this chapter, I would explore a similar trajectory in relation to the practices of music listening.

On this note, it is important to clarify that the term 'creative listener' does not imply a listener who creates; that process is more closely linked to notions of prosumption, that is, the ways in which individuals both consume music and produce music (Ritzer et al. 2012). While prosumption cultivates many levels of creativity, this argument focuses on those listening experiences that do not necessarily require the production of new music. Rather, 'creative listening,' as I define it here, is about finding new ways to enjoy the experience of listening to music within the scope of *listening as the dominant activity*. What I mean by this is that Internet listening practices call upon the listener to be creative in their understanding of what might be considered a listening experience. Unexpected and extant listening modes challenge the Internet user to redefine the boundaries of what can be understood to be a pleasurable way of listening. This emerges as a direct result of our changing relationship to the fixed-point computing system, that is, the technologizing of ourselves in relation to the potentials of our relationships with the fixed-point computing system

The nature of creativity can be understood using a number of approaches, but I focus on creativity here as it unfolds as a social phenomenon, put forth by Csikszentmihalyi and Wolfe (2014), particularly because so many Internet listening modes are the product of co-creation. Csikszentmihalyi's approach to creativity is critical for my purposes, because for him, this phenomenon is not merely a mental process, as some approaches generally assume (Csikszentmihalyi and Wolfe 2014, 162–163), but rather, emerges from "a virtual space, or system, where an individual interacts with a cultural domain and with a social field" (100). Perhaps most crucially, however, Csikszentmihalyi points out that creativity is not ahistorical. In order for something to be original and 'new,' there must be an 'old' (162). Internet listening, for example, is creative because it is a departure from traditional and long-held rituals of consumption. This is not to suggest that traditional ways of listening did not have creative aspects. However, listeners were limited to those practices restrained by their technological housing and, additionally, older technologies were not as highly interoperable. In his work on "mash up culture," Aram Sinnreich explains that in previous generations of technology "separate communicative functions were served by distinct technologies" which meant that content could not be customised and therefore produce original behaviours of consumption. However, as Sinnreich continues, "all functions are now converging on the same digital platform [therefore] all cultural participants can create, retrieve, edit, and share ... using a single tool, an Internet-connected computer" (2010, 72). Therefore, I describe creative listening in relation to "originality, freshness of perceptions, [and] divergent thinking" (Csikszentmihalyi and Wolfe 2014, 164) that are specific to the technological capabilities and social forces at work. Through the reciprocal pathways of the human-computer interface, the creative listener can now be exposed to fresh listening schemas that break from traditional models, in which music was restricted to its originally recorded form and format.

Boundaries of the PCBM and the Creative Practice of 'Geolistening'

I define geolistening as a practice in which a listener can mediate site-specific sounds from one or more remote physical locations in real-time. This is ultimately a creative practice because the listener must actively take part in the form and quality of their listening experience—they must create their listening experience—through interaction with the personal computer system. By doing so, the listener draws upon imagined intersections and it is the user that navigates the listening encounter according to their mood and technical capability, thus creating, as-they-go, new listening models that play with psycho-spatial boundaries. I will thus suggest that geolistening is a technic which orients the somatic in time and place, and thus technologizes the listening experience through virtual projections, simulations, encounters and an otherwise-beingness.

To map this idea, I take the example of the "You Are Listening to Los Angeles" streaming service. "You Are Listening To Los Angeles" (referred to from here as YALTLA) is a streaming service that picks up radio-transmissions from emergency service scanners (police, fire, ambulance, port authority, etc.) from a variety of cities and synthesizes them with one's choice of playlist. There are also options to listen to air traffic control feeds from various airports, real-time audio from online gaming feeds such as *Call of Duty*, and even a stream of archived 'JFK' speeches picked up from the University of Virginia. The listener can choose which city or feed they would like to listen from a dropdown menu and the stream will immediately transmit that feed in real-time. Traditional ontological boundaries framing time, place and being-ness are redrawn as the individual specularizes bodily boundaries which extend back in time, throughout virtual worlds, and into liminal spaces. We might understand this as somatechnology through the reference of Mark Hansen's work in *Bodies in Code*, in which he submits that:

> technologies work to expand the body's motile, tactile, and visual interface with the environment; to do so they call upon—and ultimately, refunctionalize—the bodies role as an 'invariant' ... digital technologies lend support to a phenomenological account of embodiment and expose the technical element that has always inhabited and mediated our embodied coupling with the world. (2006, 26)

Geolistening is a crafting, in that the user both takes and is taken into, and projects and is projected onto. The practice then is not a process of one object acting upon another, because this kind of listening practice can only emerge in the meeting of the two systems (human and machine) folding into one process.

Another capability of this particular creative listening practice is the production of, and intersection with, virtual tourism which redefines listening pleasure in terms relating to travel, journey, adventure and other modes of moving through space (and though travel and tourism are two very distinct experiences/constructs that distinction is less important to this discussion for the time being). However, 'geolistening' is a *development* which has emerged over several generations between the coupling of mobile technologies and listeners. Music and travel emerge in an historical context through which geolistening can be framed and personal mobile media has redrawn listening pleasure since at least the 1970s with the emergence of the cassette player. Since this time, the use of mobile media has become ultimately tied to the travel experience as a subcultural practice. Mobile media devices enabled a reconfiguration between the lines of affect, movement, discovery, space, travel and modes of listening (Bull 2014, 233–234).

However, now, with the emergence of geolistening in virtual modes, relations between music, space and travel become redefined again in new and creative fashions. For example, the YALTLA "Button Map" function remediates and thus redefines the traditional travel-music relationship, largely I argue, as a process of dissolution of traditional corporeal limitations which uses music and other aural figurations as the navigational system. The Button Map is an interactive map which can be zoomed in or zoomed out and which displays flagged listening stations marked by icons. The user can press on an icon and be taken directly to that the city's feed. In some instances, this feed will be a police scanner, but there are also icons for airport traffic control feeds, a bowling alley in Reno, and the "Deep Thought" channel which is placed in Lhasa, Tibet. (The "Deep Thought" channel is different from the other feeds in that it is not a real-time application but recordings taken from archived speeches from notable thinkers such as Aldous Huxley and Jiddu Krishnamurti). The icons act as *portals*. By interacting with this YALTLA map and streaming the different soundscapes from either physical locations or conceptual locations, such as in the case of "Deep Thought," the virtual tourist can also go places the physical tourist cannot at speeds the physical tourist cannot. In the mid-1990s, Paul Virilio wrote that, "When cosmic imagery is completely digitalised in the next century by computer processors, cybernauts will be able to travel in their armchairs as simple televiewers discovering a surrogate world that will have emerged from information energy" (1995, 154–115). I suggest that the geolistening platform is a creative and imaginative audio-visual experience in which the user becomes a kind of cybernaut, to borrow Virilio's term, who mediates their travel experience using the practice of music listening.

Perhaps one of the most provocative of the YALTLA portals is The Bloop, a site represented by a green icon floating in the middle of the Pacific Ocean south-west of Guam, where the phenomenon was originally discovered. "The Bloop" is a sound that was captured in 1997 by the U.S. National Oceanic and Atmospheric Association from the Ocean Explorer's acoustics program. When the sound was first recorded its source was a mystery, which resulted in wild speculation about massive unknown deep sea creatures, hence the title of the portal: "Cthulhu Fhtagn" after H.P. Lovecraft's fictional cosmic sea entity. In 2012, NOAA reported that they believed the sound to have emanated from an icequake (Steadman 2012) which are generated from large icebergs as they crack. What is pertinent here is not where the sound comes from but where the sound 'takes' the listener—that is, to the bottom of the ocean. Again, this is a marked distinction from traditional forms of tourism (one cannot generally visit the bottom of the ocean), however, in the event of geolistening the virtual tourist is 'pulled into' the ocean by structures of spatial metaphors—it feels possible to go to the bottom of the sea even though consciously we know this is not possible. If, referring back to Copland, the "freely imaginative mind is at the core of all vital ... music listening" (1952, 7), then the 'personal computer bodymind' field of relations produces a new dimension which redraws imaginative boundaries in the listening practice. In this example, the user is drawn 'deeper and deeper' into the multimedia experience, and thus, into the wonder of abstract space which has no limits. In this sense, the PCBM field of relations radically reshapes the experience of listening because the computer modifies the boundaries of the body and therefore re-establishes where and how listening practices emerge and make sense to us. The computer belies the traditional 'edges' of human being-ness. The PCBM coupling gifts these extended imaginative and creative potentials.

Creativity and Collective Consciousness with the PCBM

Extending somewhat from the argument initiated in Chap. 3 (Live Music) based on the entanglement between music listening pleasure and visual culture, I continue here to suggest that new computer-bodymind relationships have rearticulated the synthesis between music and viewing, in that one sensory/mediatic experience informs and transforms the other.

The PCBM *sews together* the affective potentials of watching and listening because of the affordances of the technology which, in the digital context, is differentiated by the unadulterated exchange of user-generated content. This field of exchange, through which users come together through music viewing/listening practices, helps to form a kind of collective consciousness that vitiates the perimeters between Self and Other, and between individualized listening and group listening.

So as not to get bogged down with Durkheim in a muddy digression on the nature of consciousness, I approach collective consciousness here simply as process of flux "in which individuals who imagine themselves as part of any given community, and offer material exchanges to that group," thus shape and form a "collective representation from a synthesis of their individual representations" (Wiley 1988, 257). I want to suggest that the emerging collective consciousness brought about through user-generated communities mediated through the PCBM marks an 'intercorporeality' between bodymind and machine, which we might understand following Gail Weiss' acknowledgement that, "To describe embodiment as intercorporeality is to emphasize that the experience of being embodied is never a private affair, but is always already mediated by our continual interactions with other human and nonhuman bodies" (1999, 5). The premise of which gives rise to technicities of listening, and the bringing forth of new pleasures in the entanglements between bodymind and machine, that is, the PCBM. Margrit Shildrick reminds us that this particular train of thought "echoes the Deleuzian approach in that what is of consequence is not the 'content' of any particular entity—organic or nonorganic—but the provisional instantiations that emerge from the interconnectivity of multiple forces" (2015, 18). The PCBM coupling thus extends the coming together of music and sound, and the pleasure one has of controlling what kinds of images and music listening strategies might come together to enact certain experiences of that music.

To explore this assertion in greater detail, I draw upon the example of vaporwave, a style of music which emerged in the early 2010s. Vaporwave is often referred to as an 'Internet-only' genre because it emerged solely on and through digital platforms (Wolfenstein 2015), mainly Reddit, YouTube, Bandcamp, Tumblr and Soundcloud. I argue that the vaporwave listening experience is a specifically creative and imaginative listening experience because it is contingent upon collective and social exchanges which are used to 'build' and 'build upon' the genre and which can only be accessed through the somatechné of the PCBM. Listeners do not only listen; they watch, they interact on social media, and they enter into dialogue—all orbiting around the vaporwave listening experience as a moment of collective consciousness.

Musically, vaporwave is an electronic music comprising sonic rudiments (tones, beats, timbres and so forth) hybridized from a wide variety of 'background musics,' largely muzak®,[1] as well new age ambience, 'on-hold' music, cocktail jazz, and other "corporate sonic ephemera of the 80s and 90s" (Ward 2014). The music itself is then chopped in with slowed-down repetitive samples and drowned heavy reverb (Wolfenstein 2015; Galil 2013). The word vaporwave may derive partly from the term vapor*ware*, a word used by the technology industry to describe a product that is marketed to the public but was never actually released or that never even existed (Calore 2009). The term has other implications too. For example, music commentator Adam Harper writes that, "The name 'vaporwave' itself is reminiscent of a famous passage from Karl Marx's and Freidrich Engels' *Communist Manifesto*, 'all that is solid melts into air,' referring to the constant change society is subjected to under bourgeois capitalism" (2015). In fact, while Harper's quote is correct (Marx and Engels [1848] 2007, 11), the word 'melts' is an inaccurate translation of the original text. The original text uses the word *verdampft* which means 'evaporates' or even 'vaporises' rather than 'melts', which resonates even more sharply with the term 'vapour'.[2] The name invokes that which is lost in the tides of technology and consumerism and, I would add, those objects created by the illusion of marketing (see Glitsos 2016). In line with this theme, vaporwave is a project in which the artists aim to explore "soulless techno-corporatism, with accompanying videos that draw on early Internet imagery: glitch graphics, late-90s web design, and cyberpunk aesthetics" (Harper 2015). The vaporwave visual aesthetic is thus integral to the totality of the vaporwave listening experience. The sound of vaporwave does not exist independently from its visual accoutrements, to the extent that some Redditors suggest that vaporwave is first and foremost a visual medium with sound as a secondary layer (see subreddit VaporwaveAesthetics). In this way, vaporwave defies traditional music conventions that typically privilege the music over the visual form and provides a new mode of music listening which brings forth all

[1] The word muzak is often used to refer to a generic type of music, but the term muzak® denotes a registered trademark belonging to the company Muzak® Holdings, which was sold in 2011 to the Mood Media company. It is the latter term I use in this chapter.

[2] I thank Professor Peter Beilharz for this detail.

the potentials of visual symbolism as structure of social exchange and cultural critique. Most importantly, this particular creative form of listening is a technicity which serves to connect the listener with a collective consciousness, through which emerges a dialogue around late-stage capitalism and technocracy, a dialogue accessed through these networks of personal computer-bodyminds.

As I mentioned, vaporwave draws largely from the genre of muzak®, which is an easy-listening style of music often played in shopping centres or, in more Americanised phraseology, 'the mall'. For vaporwave listeners, this aesthetic provokes:

> That feeling of being a bystander at the arcade waiting for your turn (nanosmusics 2015).

> Slightly-dystopian "What if"? style of music […] makes us wish we could be in that non-existent period of time just so we could walk down a gigantic mall with our friends with our Walkman and brightly-patterned clothes (ChiptuneGhosts 2015).

> […] the busy yet empty mall analogy (Kidneybot 2015).

> [a] futuristic/dystopian feel, or 'world that never was' vibe (lifeanddecay 2015)

> [a sense of the] hyper-real, consumer-driven world of the early 21st century. For many of us, this was a relationship that was built on the technological utopianism of '90s culture, but one that has now been challenged by the global recession and by the resultant collapse of the ideals of the free market, corporate power, and technological progress that so defined the '90s. (DasModernist 2015)

This language suggests vaporwave listening as an enterprise in collective emotional exploration; the exploration of worlds and states of being that can be envisioned but never actualized. The genre of vaporwave is subtitled as, 'music optimised for abandoned malls,' because it aims to produce the effect of a postconsumer environment where the relics of mall culture survive and the human subject exists in a transfixed purchasing trance. These notions echo the Derridian concept of "hauntology," which comes from the French *hantologie* following Derrida's use of the word in his text *Spectres of Marx* in 1993. As Colin Davis explains, the French word *hantol-*

ogie "supplants its near-homonym" *ontologie*, thereby "replacing the priority of being and presence with the figure of the ghost as that which is neither present nor absent, neither dead nor alive" (2005, 373). Vaporwave's 'hauntological' aesthetic calls forth and explores listeners' experience of the future that was advertised but never delivered (as in the term vaporware). In this sense, I frame the PCBM vaporwave phenomenon the pursuit of an active connection to these cultural experiences which permeate the collective consciousness

Derrida's *hantologie* is also manifest in the 'deadmalls.com' phenomenon, which intersects and interacts thematically with vaporwave's preoccupation with the abandoned mall and the sense of decaying late-stage capitalism in the collective consciousness. The Dead Malls website is a collection of images and stories of the malls which have been left abandoned and neglected across America. The images of rotting suburban spaces and the manifest legacies of the global financial crises are haunting and ghostly. These 'dead malls' exemplify vaporwave's characterisation of a future that is here but also never arrived. In Adam Trainer's chapter "From Hypnogia to Distroid" (2016), he writes that, "What sets vaporwave apart as an aesthetic system is the lack of direct comment taking place in the music itself ... vaporwave presents a simulation of both the empty vacuousness and the boundless promise of a postglobalised consumer landscape" (421). This forms an approach to culture brought about by the collective consciousness delivered through and building from the personal computer bodymind relationscape, that is, in Erin Manning and Brian Massumi's philosophical inquiry, "the malleability of concepts that move, the expressivity of thoughts as they become feelings, the ontogenetic potential of ideas as they become articulations" (2012, 5). Vaporwave listeners therefore explore, and are positioned within, a very particular affective schema specific to the framework of consumer relations in the digital age in which the listener is haunted by the spectres of better, or at least different, futures and which are brought forth through the PCMB somatechnic.

Further, the symbol of the mall plays a very significant role in cultivating the listening experience because of the tightly-bound relationship that has been constructed between the mall and the contemporary subject and which is subsequently articulated through the vaporwave aesthetic as it is accessed through the personal computer (through Reddit threads, YouTube videos and comments, Tumblr dialogue and so forth). In Jonathan Sterne's words, the mall is an "icon" for hyper-consumerism and the soundtrack of this space is undoubtedly muzak® (1997, 22). For Sterne, the muzak® disseminated throughout The Mall of America, the largest mall in the U.S., constitutes its very architecture: "Rather than simply filling up an empty space, the music becomes part of the consistency of that space. The sound becomes a presence, and as that presence it becomes an essential part of the building's infrastructure" (1997, 23). Since the emergence of muzak® in the 1930s (Muzak® Holdings was founded in 1934), music has become implicated in corporate strategies as an accomplice to consumer manipulation and workforce management, helping consumers shop for longer and encouraging personnel to work more productively (Jones and Schumacher 1992, 156). In this instance, music is therefore stripped of any sense of artistic individuality or potentialities (Lanza 1995, 67–68). Instead, music itself becomes absorbed into the matrix of mall culture. In Sterne's critique of The Mall of America:

> The economics and social organisation of programmed music presumes and exists on top of a whole culture and economy of recorded music. In other words, programmed music presumes that music has already become a thing—a commodity. (1997, 24)

The vaporwave phenomenon forms a parody of the commodification of contemporary music as well as the broader commodification of culture itself. However, what interests me most here is that by drawing upon and repurposing muzak®, vaporwave 'works at problems' through the collective consciousness *by way* of the PCMB.

However, as with all models of change and flux, there are resistances and anxieties brought about by new syntheses between human and machine. Vaporwave both explores and expresses the ever-increasing porous nature between the flesh and what Haraway would call "the cybernetic fold" (1987). That is, there is a self-awareness in the vaporwave narrative and aesthetic which calls attention to the inescapable entanglement between not only human and technology, but the human, technology and capitalism. Thus, this brings about an unease and even a mistrust for that which permeates our bodily borders, that which can seep into our lives, and that which can holdfast the very contents of our memory. For example, vaporwave's site of interest is the "virtual plaza," in Adam Harper's words, the virtual plaza is "literal and metaphorical, real and imaginary—the public space that is the nexus of infinite social, cultural and financial transactions and the scene of their greatest activity and spectacle" (2012, para. 8). The virtual plaza is a post-consumer plaza. A shopping mall for the end of the world, so to speak, in which all individuals are subsumed into the role of consumption and all activity in everyday life falls under that rubric. There is no more 'consumer' and 'non-consumer' because *all* subjects are interpellated into the architecture of consumption. For music commentator Robert Ham, vaporwave therefore constitutes:

> A decidedly dystopian genre [...] glistening beats for a hyper-glassine sound that would feel equally appropriate playing underneath a modern cut of *Blade Runner* or a first-person shooter zombie attack video game. (2012, n.p.)

In this sense, vaporwave invokes very similar experiences as these popular science fiction texts that explore and consider terrifying possibilities in which the subject has already been transformed into a 'mindless drone' through the coupling of the bodymind-machine, similarly to the narrative which emerges in *The Matrix* trilogy. For example, Jon Stratton reads *The Matrix* trilogy as an exploration of a dystopia, with a "twist" (2006, 29), where the "machines provide an albeit illusory world which is better than the 1999 peak of human civilisation" (30). Stratton goes on to quote James Berger who explains that while modernity was "'preoccupied by a sense of crisis,'" in more contemporary contexts, "'This sense of crisis has not disappeared, [...] it exists together with another sense, that the conclusive catastrophe has already occurred'" (Berger 1999, as cited in Stratton 2006, 36). Berger suggests that we might not even know exactly when this catastrophe occurred but "'the ceaseless activity of our time...is only a complex form of stasis'" (Berger 1999 as cited in Stratton 2006, 36). In the first *Matrix* film, for example, Morpheus explains to Neo that humans are not sure of the exact date and that the details of when and how the machines took over the human race is not known ("although it was us that scorched the Sun" as Morpheus confesses). Humans 'go about their business' completely oblivious to the fact that they are in fact in a dream state being drained of their life-force. In a parallel sense, vaporwave constructs and represents an unease emerging in the collective consciousness which manifests in the very nature of the genre— that the individual is unwittingly subjugated to the realm of the capitalist dystopia in which all transactions take place within the unreal environment of the 'virtual plaza.'

In another sense, the vaporwave project digs up that which capitalism discards, and brings it to the fore: old VHS tapes, bad haircuts, the grating tones of corporate instructional videos, and so forth. In relation to a personal computer bodymind somatechnic, I suggest this is a technicity of cultural purging, which follows on from the notion of collective consciousness. By marking this practice through the listening experience, vaporwave serves as a type of cultural purge of old and buried material. For one Redditor, vaporwave is "the most cathartic" music s/he has ever heard (joshuatx 2015). Applying Eldritch Priest's work on experimental music is pertinent here. For Priest, muzak® is "the public shame of good taste" (2013, n.p.). The use of shame as a collective emotion is critical here. As Julia Kristeva famously explores in *Power of Horror* (1982), shame is representative of that which is abject, those materials that culture seeks to rid itself of (8). And yet, vaporwave takes this shame as its launching point and with this 'shame' creates new sounds. As Priest explains, recent experimental compositions have repurposed many of those sounds which have been thrown away, made disposable, or discarded as waste material (or as he calls it, that which has been made into "shit"). Experimental music calls forth these obsolete sounds in order to "fertilise the wide field of listening with a farrago of attentional spores that sprout gnarled shoots of interest to see new aesthetic sensibilities" (2013, n.p.). Following along these lines, I suggest that vaporwave, as an experimental form of music, repurposes those artefacts that our culture has repressed or disposed. Vaporwave seeks out that which has been lost in order to chart those lost affects. Priest explains that those "habits of inattention developed around the use of ubiquitous audio media forms" has helped to create music that "replicates and warps the drifts and digressions that constitute those habits" (n.p). Thus, the vaporwave project utilizes the PCBM field of relations to interrogate the limits and potentials of our collective consciousness.

The Era of the Algorithm: Listening and Discovery with the PCBM

Lastly, I wish to examine the way that the personal computer bodymind field of relations enacts a new moment in imaginative processes around musical discovery. I propose the way in which this happens is through the reciprocal and conjoint processes of listening online that exposes one's listening patterns to algorithmic technologies. Though, I would concede that not all modes of discovering new music are necessarily bound to practices of listening, certainly for my purposes here I do focus on those practices in which the discovery of new music is contingent upon listening. While many traditional modes of musical discovery from the pre-MP3 era have diminished (stalking through record libraries, being immersed in headphones at CD store listening booths, taping songs off the radio and so forth), there are digital discovery strategies which have emerged and reimagined the relationship between music listening and music discovery. Most significantly, we find ourselves in the 'era of the algorithm,' and I will expound on the significance of this context in relation to musical discovery in order to argue that the new model of discovery/listening is *not* so much led by the individual but directed, shaped and qualified by the personal computer bodymind field of relations. In a way, this field of relations is a unified consciousness which draws upon the power of the algorithm to explore new affective potentials.

The unified field of relations that constitutes the personal computer bodymind effectively acts as an extended consciousness connecting user and the potentials of discovery through the *experience of listening*. That is, the system relies on the user to listen in order to generate discovery, to be open to creative suggestions brought about by the algorithmic system. The more the user listens, the more personalized and tailored this experience becomes (Celma and Lamere 2011). One listens in order to listen more. This occurs within the unified field of relations—this somatechnic—which extends the self into the digital fold and its unimaginable potential. Let us explore this assertion using the example of the complex programming models of music recommendation systems.

There are a range of algorithmic models (creatively) created by programmers which serve to open the listener up to new musical discovery processes. As explained by Òscar Celma and Paul Lamere, these are generally based on programs which are used to derive similarities between musical texts and help users 'discover music' based on those similarities. Three of these systems are known as the Usage-Based Recommender, the Social-Based Recommender, and the Content-Based Recommender (2011, 60–61). In terms of the Usage-Based Recommender, music suggestions are generated by filtering the users' listening habits and then analyzing the usage patterns. The "collaborative filtering" works by "building a matrix" of the interaction between users and items—plays, ratings, page views, and so on (60). The Social-Based Recommender system works by computing "similarity among items through web mining techniques, or exploiting social tagging information" (60). The authors explain the complexity of this system below:

> Web mining techniques aim at discovering interesting information from the analysis and usage of web content. Similarity among artists is computed based on, for instance, co-occurrence analysis in web pages, songs played in the same session log, or the text analysis of album reviews. Social tagging aims at annotating web content using tags. A bottom-up classification emerges when all the annotations (tags) from a tagging community are merged. (Celma and Lamere 2011, 60)

Finally, the Content-Based Recommender system "extracts features directly from the music and uses these features to determine item similarity" (61). In this approach, audio similarity must be computed, usually "based on low-level timbre descriptors" (or what is called Mel frequency cepstral coefficients) (61). Celma and Lamere explain that this approach can also improve genre classification by modelling "the audio signal using a statistical distribution of the audio features on short-time audio segments. Audio features are then aggregated using simple statistics (for example, mean and variance), or modeled as a Gaussian Mixture Model (GMM)" (61). However, these systems functioning alone typically have some limitations, so many recommendation systems will now deploy a synthesis of these techniques, known as the Hybrid Recommender system (61).

In giving a cursory overview of some simple approaches to recommendation systems through which listeners pursue musical discovery, my aim is to point towards the complexity of system interactions which is at play in the process of digital music discovery, and which are bound to the listener's pre-figured listening experiences. I wish to emphasize that the technology borrows from the human and the human borrows from the technology. Ultimately, this is a process of interactionism through which the listener finds joy, excitement, pleasure and a wealth of affective sites on the listening spectrum. However, and most importantly, there is a line where the borrowing becomes a *breaching*—that is, the exchange between the personal computer and the bodymind conflates the binary between them and forms the unified field of relations of the personal computer bodymind. The listener invests a great part of themselves within the 'hold' of the personal computer, in a sense, investing fragments of their beingness (what they feel, think, watch, listen to, enjoy, dislike and so forth), and the connected technologies are utilizing those fragments as a bridge, almost reaching *inside* the listener, to extract the information it needs in order to feed that listener's experience.

Perhaps even more exemplary of this assertion is the programming work done by Shan et al, in which they propose a "generic framework for emotion-based music recommendation by affinity discovery from film music" using a "modified Mixed Media Graph algorithm and Music Affinity to discover the relationship between music features and emotions from film music" (2009, 7666). Put simply, this research aims to find relationships between music and emotion in film experiences. By collating this data, the final aim is to be able to "recommend music based on emotion" (7673). The way this works is, in very basic terms (the profoundly complex version of which can be found in their research article), is that the 'recommended list' of music for the listener emerges from a ranking generated by analyzing the matrix of interactions which forms the relationships between 'film segment emotion detection' and the correlating 'music feature extraction,' which is then measured against 'database music features' (7667). This process exemplifies the crux of 'somatechnics,' which we should recall is a "term, derived from the Greek sôma (body) and τεχνε (craftsmanship), [that] supplants the logic of the 'and,' suggesting that technés are ... the dynamic means in and through which corporealities are crafted, that is, continuously engendered in relation to others and the world" (Sullivan and Murray 2009, 3). For example, the case of 'emotion based music recommendation by affinity discovery from film music,' is a process of the "machinization of humans and the humanization of machines" such that each system absorbs and rearticulates the other (Yan Zheng 2017, 112). Rosalind Picard coined this kind of process "affective computing" to describe "'computing that relates to, arises from, or influences emotions'" (Picard 1995 as cited in Yan Zheng 2017, 111). Caroline Yan Zheng explains that:

> For affective computing to 'machinize humans' is an activity of collecting and analyzing human data and recognizing behavioral patterns over time. These behavioral patterns are then modelled and installed on machines as artificial intelligence. To perfect this human machine relationship, especially on an individual level, motivates us to gain knowledge about our emotion selves. (Yan Zheng 2017, 113)

This is the kind of emotional reach we see unfold in the discoverability of new music and we can understand this as creative because the listener engages in an imaginative, open and exploratory relationship within the PCBM field of relations.

The listener is open to being transformed by the potentials of an algorithmic model constructed by both machine and other humans, and coalescing on everyday streaming platforms. Additionally, listeners tailor their listening experience in lieu of their relationship to the algorithm. I offer this anecdote from one Spotify user who posted their experience on Reddit.com:

> It's been my experience that making playlists radically alter the direction that my future discover playlists take. As an example, I listen mostly to hip hop and not EDM electronica (IDM, trip hop, future bass, etc.). If I made a twenty-song indie playlist of indie stuff I listened to ten years ago, and my discover playlist was all indie for weeks. To rectify that, I made two playlists, one hip hop and one electronica. It was fixed the following discovery update. Afterward, if my playlist ever starts to veer away from what I actually want to listen to, I create a new small playlist or two, OR add songs to previously existing playlists, and the course is corrected. (LookingForVheissu 2018)

The creative potential of each system works as one aspect of the entire PCBM unit, in order to seek out new affective dimensions in the digital listening schematic. I offer Yan Zheng here to emphasize that:

> The boundaries between body, space, physical and digital have merged. Here, technology and materiality become one; coding and making interweave; the mediating power of computation couples with that from the aesthetics and sensual qualities of artefacts. However, this hybridity also adds layers of complexity to already sophisticated affective computing technologies. (Yan Zheng 2017, 116)

To end, algorithmically produced modes of discovery are contingent upon the listening practices of selves and others. The more one listens through the PCBM field of relations, the more one discovers musical experiences through that PCBM field of relations. The ability to be open, exploratory, and imaginative is vital in this process, much as it is in the very experience of music listening itself. It is my contention that these processes fold over and into one another in new entanglements which offer the exciting potentials of algorithmic models as well as the horizon of possibility that is the personal computer bodymind somatechnic.

Conclusion

In this chapter I have emphasized that the synthesis between the listener and the personal computer suggests more than a mere pairing of relations. Rather, these two highly complex systems are sewn together, forming a somatechnic in the most exemplary sense of the concept. I have suggested, in fact, that we can read this somatechnic in terms of a 'personal computer bodymind field of relations' which extends the bodymind by twisting itself into the Mobius bend that delimits inner and outer worlds. The personal computer bodymind field of relations is a shift in a consciousness that thus brings about changes in collectivities and creativities. Robert Rawdon Wilson declares that:

> Even glasses modify consciousness. It is not merely that I can now see better, but also that an aspect of my being has been put behind me, but never out of mind as well. My ocular prosthesis elevates me to a higher plane of fulfilment, towards a more ideal conception of myself. (1995, 239)

Entire music collections are maintained on computers (and often backed up on hard drives or on the cloud, but again, those can only be accessed by connecting to a computer) in a way that extends the bodymind far beyond its organic capability, much like a set of eyeglasses. The computer extends the bodily being-ness. Streaming services, YouTube accounts, Spotify playlists, and other services maintained by fast and stable Internet connection are all at the mercy of those structures of communication between human and computer. Thus, the bodymind forms a radical coupling with that portal—the computer—which regulates, mediates, maintains and modifies all those practices of music listening that the contemporary experience is contingent upon. In order to engage with music, the listener can (and sometimes must) capitulate to the extant listening practices that have reshaped, and continue to transfigure, traditional music experience. The listener brings a creativity, an openness, a process of reflection to how, where, and in which formats one might experience listening.

I have situated the creative listener in terms of a somatechnical framework because, as I have argued, these listening practices are dependent on the amalgamation between two highly complex, semi-closed systems—the human bodymind vis-à-vis the computing apparatus. I call both of these systems 'semi-closed' because, while both the bodymind and the computer maintain their integrity, both are always open to each through fissures and circuits, as well as being amenable to the wider discursive structures that both engulf and produce them. Needless to say, the model of the creative listener is antithetical to the mainstream ideal which draws clean lines between the material body of the user and their 'virtual' activity as it is mediated through the screen interface. In fact, instead, I suggest the computer very much reaches out and reshapes our creative potentials which determine the pathways of our music listening pleasures and experiences, that is, how and why we listen.

References

Bull, Michael. 2014. Sensory Media: Virtual Worlds and the Training of Perception. In *A Cultural History of the Senses*, ed. Constance Classen, J.P. Toner, Richard Newhauser, Herman Roodenburg, Anne Vila, and David Howes, 219–241. London: Bloomsbury.

Calore, Michael. 2009. Vaporware 2009: Inhale the Fail. *Wired*, December 21. http://www.wired.com/2009/12/vaporware-2009-inhale-the-fail/

Celma, Òscar, and Paul Lamere. 2011. If You Like Radiohead, You Might Like This Article. *AI Magazine* 32 (3): 57–66.

ChiptuneGhosts. 2015. Why I Think We Like Vaporwave. *Reddit.com*. https://redd.it/2zul91

Copland, Aaron. 1952. *Music and Imagination*. Cambridge, MA: Harvard University Press.

Csikszentmihalyi, Mihaly, and Rustin Wolfe. 2014. New Conceptions and Research Approaches to Creativity: Implications of a Systems Perspective for Creativity in Education. In *The Systems Model of Creativity: The Collected Works of Mihaly Csikszentmihalyi*, ed. Mihaly Csikszentmihalyi, 161–184. Dordrecht: Springer Netherlands.

DasModernist. 2015. Why I Think We Like Vaporwave. *Reddit.com*. https://www.reddit.com/r/Vaporwave/comments/2zul91/why_i_think_we_like_vaporwave/cpngvoz

Davis, Colin. 2005. Hauntology, Spectres and Phantoms. *French Studies* 59 (3): 373–379.

Durkin, Erin. 2018. Hacker Sentenced to Prison for Role in Jennifer Lawrence Nude Photo Theft. *The Guardian*, August 30. https://www.theguardian.com/technology/2018/aug/29/nude-photo-hacker-prison-sentence-jennifer-lawrence-victims

Galil, L. 2013. Vaporwave and the Observer Effect. *Chicago Reader*, February 19. http://www.chicagoreader.com/chicago/vaporwave-spf420-chaz-allen-metallic-ghosts-prismcorp-veracom/Content?oid=8831558

Glitsos, Laura. 2016. Ways of Feeling: The Transformation of Emotional Experience in Music Listening in the Context of Digitisation. PhD diss., Curtin University.

Ham, Robert. 2012. Gatekeeper Exo: AllMusic Review by Robert Ham. *Allmusic*, July 17. http://www.allmusic.com/album/exo-mw0002387718

Hansen, Mark. 2006. *Bodies in Code*. New York: Routledge.

Haraway, Donna. 1987. A Manifesto for Cyborgs: Science, Technology, and Socialist Feminism in the 1980s. *Australian Feminist Studies* 2 (4): 1–42.

Hargreaves, David, Jonathan James Hargreaves, and Adrian C. North. 2011. Imagination and Creativity in Music Listening. In *Musical Imaginations: Multidisciplinary Perspectives On Creativity, Performance and Perception*, ed. David Hargreaves, Dorothy Miell, and Raymond MacDonald, 156–172. Oxford: Oxford University Press.

Harper, Adam. 2012. Comment: Vaporwave and the Pop-Art of the Virtual Plaza. *Dummy Mag*, July 12. https://www.dummymag.com/features/adam-harper-vaporwave

———. 2015. Pattern Recognition Vol. 8.5: The Year in Vaporwave. *Electronic Beats*. http://www.electronicbeats.net/vol-8-5-the-year-in-vaporwave/

Heidegger, Martin. 1977. *The Question Concerning Technology, and Other Essays*. New York: Harper & Row.

Jones, Simon, and Thomas Schumacher. 1992. Muzak: On Functional Music and Power. *Critical Studies in Mass Communication* 9 (2): 156–169.

joshuatx. 2015. Has 2814 Completely Ruined Vaporwave? *Reddit.com*. https://www.reddit.com/r/Vaporwave/comments/3kw49p/has_2814_completely_ruined_vaporwave/cv28avl

Kidneybot. 2015. Why I Think We Like Vaporwave. *Reddit.com*. https://www.reddit.com/r/Vaporwave/comments/2zul91/why_i_think_we_like_vaporwave/cpmgcw4

Kristeva, Julia. 1982. *Powers of Horror: An Essay on Abjection*. New York: Columbia University Press.

Lanza, Joseph. 1995. *Elevator Music: A Surreal History of Muzak, Easy-Listening and Other Moodsong*. London: Quartet.

lifeanddecay. 2015. What Is Your Definition of Vaporwave. *Reddit.com*. https://www.reddit.com/r/Vaporwave/comments/3akhwo/what_is_your_definition_of_vaporwave/

LookingForVheissu. 2018. Regarding the Discovery Complaints. *Reddit.com*. https://redd.it/9214kr

Manning, Erin, and Brian Massumi. 2012. *Relationscapes: Movement, Art, Philosophy*. Cambridge, MA: MIT Press.

Marx, Karl and Engels Freidrich. [1848] 2007. *Communist Manifesto*. Radford, VA: Wilder Publications.

McLuhan, Marshall, and Quentin Fiore. 1967. *The Medium is the Massage*. New York: Bantam.

nanosmusics. 2015. Why I think We Like Vaporwave. *Reddit.com*. https://www.reddit.com/r/Vaporwave/comments/2zul91/why_i_think_we_like_vaporwave/cpngvoz

Priest, Eldricht. 2013. *Boring Formless Nonsense: Experimental Music and the Aesthetics of Failure*. New York: Bloomsbury.
Prior, Nick. 2018. *Popular Music, Digital Technology and Society*. London: Sage.
Rawdon Wilson, Robert. 1995. Cyber(body)parts: Prosthetic Consciousness. In *Cyberspace/Cyberbodies/Cyberpunk: Cultures of Technological Embodiment*, ed. Mike Featherstone and Roger Burrows, 239–260. London: Sage.
Ritzer, George, Paul Dean, and Nathan Jurgenson. 2012. The Coming Age of the Prosumer. *American Behavioral Scientist* 56 (4): 379–398.
Shan, Man-Kwan, Fang-Fei Kuo, Meng-Fen Chiang, and Suh-Yin Lee. 2009. Emotion-based Music Recommendation by Affinity from Film Music. *Expert Systems with Applications* 36 (4): 7666–7674.
Shildrick, Margrit. 2015. 'Why Should Our Bodies End at the Skin?:' Embodiment, Boundaries, and Somatechnics. *Hypatia* 30 (1): 13–29.
Sinnreich, Aram. 2010. *Mashed Up: Music, Technology, and the Rise of Configurable Culture*. Amherst: University of Massachusetts Press.
Steadman, Ian. 2012. The Bloop Mystery Has Been Solved: It Was Never a Giant Sea Monster. *Wired*, November 29. http://www.wired.co.uk/news/archive/2012-11/29/bloop-mystery-not-solved-sort-of
Sterne, Jonathan. 1997. Sounds like the Mall of America: Programmed Music and the Architectonics of Commercial Space. *Ethnomusicology* 41 (1): 22–50.
Stratton, Jon. 2006. 'So Tonight I'm Gonna Party Like It's 1999': Looking Forward to the Matrix. In *The Matrix in Theory*, ed. M. Diocaretz and S. Herbrechter, 27–51. Amsterdam: Rodopi.
Sullivan, Nikki, and Samantha Murray, eds. 2009. *Somatechnics: Queering the Technologisation of Bodies*. Surrey: Ashgate.
Trainer, Adam. 2016. From Hypnogia to Distroid: Postironic Musical Renderings of Personal Memory. In *The Oxford Handbook of Music and Virtuality*, ed. Sheila Whiteley and Shara Rambarran, 409–427. New York: Oxford University Press.
Virilio, Paul. 1995. *The Art of the Motor*. Minneapolis: University of Minnesota Press.
Ward, C. 2014. Vaporwave: Soundtrack to Austerity. *Stylus*, January 29. http://www.stylus.com/hzwtls
Weiss, Gail. 1999. *Body Images: Embodiment as Intercorporeality*. New York: Routledge.
Wiley, Norbert. 1988. The Micro-Macro Problem in Social Theory. *Sociological Theory* 6 (2): 254–261.
Wills, David. 2008. *Dorsality: Thinking Back Through Technology and Politics*. Minneapolis: University of Minnesota Press.
Wolfenstein. 2015. *Vaporwave: A Brief History*. YouTube Video, 22:36. https://www.youtube.com/watch?v=PdpP0mXOlWM
Yan Zheng, Caroline. 2017. Machinising Humans and Humanising Machines: Emotional Relationships Mediated by Technology and Material Experience. In *Digital Bodies: Creativity and Technology in the Arts and Humanities*, ed. Susan Broadhurst and Sara Price, 111–127. London: Palgrave Macmillan.

CHAPTER 6

Future Bodies, Future Music

Music is prophecy … It makes audible the new world that will gradually become visible, that will impose itself and regulate the order of things; it is not only the image of things, but the transcending of the everyday, the herald of the future
Attali (1985, 11)

Introduction

This chapter explores how emergent modes of virtual reality and mixed reality technologies change or impact the listening experience in ways that reconstitute pleasure, in line with an understanding of somatechné. I begin this chapter with two statements. The first statement is a quote from a user discussing a new virtual reality application that enables music fans to 'enter' into a virtual dance club and listen to DJ sets in real time, as the user states: "This is the future" (travtrigs 2018). The other is drawn from the movie *The Cable Guy* (1996), in which Jim Carey, in his role as "Chip" Douglas, is madly proselytizing on the ever-incredible connections afforded by new technologies, shouting atop a satellite dish that:

> Soon every American home will integrate their television, phone and computer. You'll be able to visit the Louvre on one channel or watch female wrestling on another. You can do your shopping at home or play Mortal Kombat with a friend from Vietnam. There's no end to the possibilities! (Apatow et al. 1996)

L. Glitsos, *Somatechnics and Popular Music in Digital Contexts*, Pop Music, Culture and Identity,
https://doi.org/10.1007/978-3-030-18122-2_6

In the 90s, this was the future and it had arrived. Of course, the paradox is that the future is always now—and it is also always arriving. According to Terry Flew's work on new media technologies, the core perplexity about new media concerns what we might define as 'new' in the first instance. Flew reminds us that the "paradox of new media is that the media technologies that we now consider to be old were once new" (2014, 2). Media is, in the words of Lisa Gitelman, "always already new" (2006). However, paradoxically, as soon as media technology is introduced it is already dated and usually predicated on something which has come before.

In this respect, how and why am I positing VR and MR technologies as 'new' and couched in a rhetoric of future music? This is especially pertinent when we consider that, as I will point out, so many practices tied to VR and MR technologies actually constitute decades old behaviors—social listening, dancing to music, sharing new tracks with friends, and so forth. This is the paradox Flew wrestles with, for example, VR and MR have hitherto unrecognized technical capabilities but, simultaneously, they also extend pre-existing frameworks *already* in place. To broach this question myself, I would echo Flew in saying that understanding what is 'new' about 'new media' practices needs to be considered in relation to cultural impact and modes of convergence that enable new "flows of content across multiple media platforms" (Jenkins 2006 as cited in Flew 2014, 5). The techniques that produce VR and MR are contingent on not only convergent technologies but convergent cultural modes which dictate future patterns based on the old ones. Thus, in this chapter, I focus on the *effects* of the technologies, less than the technologies themselves, to emphasize current ideas about the body in flux.

I reflect on several practices of music listening which have emerged in the past decade and which integrate virtual or mixed realities such as TheWaveVR and 3D mapping at live shows. By doing so, I wish to put forth an understanding of where the emphasis of music listening is located and where it may continue to venture towards. What I propose is that we see an emerging trend toward virtual and mixed reality technologies in music listening which play upon notions of 'the real' (and which will likely continue as a lasting preoccupation in one form or another). However, I do not wish to exaggerate or idealize virtual realities. Justin Grandinetti and Charles Ecenbarger advise us that, "The virtual is not transcendent, in that it does not operate on another plane of existence from the real" (2018, 4). Instead, according to Stephen Wiley's Deleuzian perspective, the virtual can be read as "completely immanent to actual bodies, things and states of affairs on the single plane of reality" (Wiley 2005, 72). That is, the virtual is always operating within and produced through the subject, as such, it is "a set of abstract relations and points structuring an object, organism or content" (70). For Grandinetti and Ecenbarger "the virtual and the real meet via actualization," which is a process that can be thought through using Wiley's notion that "virtual multiplicity is incarnated in an actual body with extension or corporeality" (Wiley 2005 as cited in Grandinetti and Ecenbarger 2018, 4). It is this basis which informs the groundwork for the arguments to follow.

Most importantly, I expound on what these VR practices might tell us about the bodymind. I offer that next-generation virtual reality and 'mixed reality' technologies reconstitute the somatechnic capacities of bodymind dynamics in listening practices in a way that foregrounds the site of the body as the 'ground zero' of all experiential exchange. Put simply, these new practices indicate that what one hears, sees and does in space ceases to matter as much as the *sense* of hearing, seeing and doing in time.

The Wave VR

'TheWaveVR' is an online application available for download from the Steam Entertainment Platform and accessible by using virtual reality headsets currently on the market, namely, HTC Vive and Oculus Rift. Once connected to TheWaveVR application, and donning the VR headset, the user can enter into a "virtual venue in the cloud" (Marcus 2016, para. 4). The virtual venues host DJ sets where, "The audience can be networked in from anywhere in the world and actually be virtually represented in this venue and interact with each other, dance and watch this concert" (para. 4). The premise of the technology is to allow "people to listen to music together in an immersive digital space" (para. 7). One of the company founders, Adam Arrigo, explains that "The goal is to bring people together but also create this new type of experience" (para. 4) but that "the secondary function is to act as an interface for people to program music" in a three-dimensional environment (para. 7). Music journalist Ezra Marcus describes his first experience with the technology, by stating:

> When I slipped on TheWaveVR's headset, which uses the HTC Vive device, I was immediately transported to a seamless, crystal-clear virtual environment. I felt completely cut off from the physical world. From my first perspective in an endless black plane that stretched in every direction, the only parts of myself I could see were my hands.

Marcus reports that, currently, 'animé-type cats' serve as avatars for the audience members experiencing the concert in VR. Though the basis of the technology is for music listening, there is some talk that users themselves may also be able to construct and perform their own DJ sets. Marcus was privy to this experience:

> In front of me, a grid of empty cubes floated in space, with a band of light passing over them on a loop. Putting my hand inside any of the boxes and pulling the trigger on the controller turned the box on; when the band of light swept across it, the box made a different sound. Different nodes on the vertical axis represented different instruments—drums, synths, guitars, and more—and the horizontal axis affected pitch. In no time at all, I'd built a rudimentary loop. (Marcus 2016, para. 9)

For Arrigo, the technology serves not only as a site which combines pleasure, music listening and social connection but suggests a re-valorization of traditional music listening practices: "VR is an opportunity is to get people's focus or attention back on music," he says, "we love Spotify, but that takes your attention away from the music ... With VR, you put this headset on and [the music] is all you can focus on—You can get people to appreciate it again" (para. 11). Arrigo's sentiment expresses a somewhat nostalgic sensibility; nonetheless, his statement demonstrates the emphasis on music listening strategies as they transform with new technology.

However, my purpose is not to evaluate the effective merits of the technology in relation to authenticity of popular music appreciation, rather, I am concerned with the way VR (and MR) technologies reconstitute the somatechnic of bodymind in a way that hitches the horizon of possibility onto listening experience. As I mention above, I argue that the somatic is produced by the technic of virtuality in a way that continues to render the body as the central site of pleasure in listening. This proposal extends Mark Hansen's work in *Bodies in Code* in which he offers that:

> The new mixed reality paradigm foregrounds the constitutive or ontological role of the body in giving birth to the world. For today's researchers and artists, virtual reality serves to highlight the body's function as, to quote phenomenologist Merleau-Ponty, an 'immediately given invariant,' a 'primary access to the world,' the 'vehicle of being in the world.' The body forms an ultimate background, an absolute here, in relation to which all perceptual experience must be oriented. (2006, 5)

Thus, the constitutive framework of listening pleasure as it relates to VR technology is the somatic *as it is produced* by and through techné. This approach gives precedence to the idea that, rather than the body 'losing' itself to technology (Hansen 2006, 2–3), these kind of "'interactive media are supporting the multisensory mechanisms of the body and are thus extending man's [sic] space for play and action'" (Fleischmann and Strauss as cited in Hansen 2006, 2–3). In following this line of thought, we might also say that the human produces the capacity of the machine as much as the machine produces those capabilities of the human.

The question, and core of this argument, however, is how this shapes the actual pleasure of listening and why. First, I would argue that this extension of the bodymind into the world of virtual relations extends human capacity as a multisensory advance into a new frontier of space-time relations, much like the experience of walking—or in this particular case listening to music—on a different planetary surface might bring about. This approach considers 'virtuality' as contingent upon the body and 'the real,' and shifts the paradigm away from immaterialism (about virtual discourses) and more towards its deployment as a technicity for extensions of bodily-consciousness. Hansen reminds us that, technicity, in this sense, is a "relation to exteriority, [and] as exteriorization, [it] is not and cannot be something merely added on to some 'natural' core of embodied life. Rather, it must be understood to be a constitutive dimension of embodiment from the start" (2006, x). Then, to follow, what pleasure of experience the bodymind 'loses' in the virtual space (that is, actualization of material objects through traditional modes of tactility), the bodymind appropriates from other forms of bodily-beingness—such as the new definitions of social bodies, new zones of interactions, and new encounters with the experience of proprioception, which I will explore now.

Somatechnics of Virtual Social Bodies

The New Social Listening Bodymind as Constituted in TheWaveVR

TheWaveVR permits and produces a new somatechnic; that is, the technology constitutes a new way of experiencing the social and socialized body through collective music listening practices. For instance, in viewing one particular WaveVR interaction which has been shared on YouTube, one can witness a user 'make a friend' while dancing and listening to the DJ set (G-Dub 2018). The user (heard giggling in the background) pulls up a virtual console using hand gestures and selects the 'SOCIAL' category. This action brings up a pink 3D heart with "add friend" written on the front of the icon. The user then takes this heart and pins it to another user's chest, or rather, pins it to their avatar, upon which the other user can choose whether or not to accept the friend request. This process functions somewhat as a remediation of the Facebook 'add friend' feature, reimagined in the virtual experience of a dance club.

The first thing that interests me about this encounter is that the avatars are not gendered by traditional markers, rather they are represented by a more neutral symbolic code (that is quite similar to cartoon cats), or what one commentator calls "genderless bubble-like characters with eyes, smiles, and little else" (Rubin 2018, para. 4). Sometimes, users might secure handles that are suggestive of traditionally gendered names (such as 'louisboy1'), but the gender identity of the user still remains ambiguous or, most often, altogether unknown. This reconfigures the normative experience of the dominant (material) somatic encounter within a traditional dance club, where behaviors and codes are contingent on (often) heteronormative models of desire, sex, and even friendship. Instead, these codes are radically rewritten and thus regenerate the relationship between sex, desire and the shared-body in the dance club social listening experience. Where, in many (although, not all) traditional club scenes, socialization practices are ultimately attached to the performance of gender (see Hesmondhalgh 2013, 109), in this new interactive zone one may 'plug into' the reconfigured political space, or what one user calls, an "all-too-rare distillation of VR's greatest promise: Sharing something beautiful, and otherwise unattainable, with another person" (Rubin 2018, para. 8). This represents a shift in how we view the capacity for the bodymind to connect in space, with music, and with other subjects. The experience of the social space becomes somewhat disengaged from traditional normative practices, especially around gender, but also around other normative markers of race, class and ethnicity because they cannot be readily identified in this arena. In some ways, though I would concede not all, this both constructs and extends a utopian sensibility which first emerged with social listening practices since at least the organization of large-scale music festivals of the 1960s, the quintessential being the discourse of 'peace and love' tied to the Woodstock Music and Arts Fair in 1969.

What is also interesting is not only the differences TheWaveVR poses to traditional club relations, but also the *similarities* of the practice to traditional club relations. For example, David Hesmondhalgh writes that in the club space, "dancers might find themselves expressing themselves more freely, and gradually relinquishing their initial self-consciousness … freed from the inhibition of their excessive self-monitoring" (2013, 109). TheWaveVR experience *builds* on this premise in order to 'open up' these fissures and gaps so that the individual may emerge through them. On this note, it is worth explaining that TheWaveVR scene replicated here is that of 'rave subculture' or 'dance subculture'. These subcultures offer TheWaveVR discursive maps which have been built from a pre-existing framework of subcultural spaces and practices. In the late 1980s and 1990s, "rave culture involved the creation of new forms of community-based on intensely corporeal and pleasurable activities and the loss of the self within the crowd" (Hesmondhalgh 2013, 111). It is my contention that TheWaveVR continues a decades-long tradition of the social dancing body as it is imagined by the dance or rave subculture yet rearticulated with new potentials brought *through* the bodymind-in-VR somatechnic.

Tripping with Friends in the Virtual Dance Club: Social Interactions as Constituted in TheWaveVR

TheWaveVR also reimagines and extends the relationship between social listening and psychedelic experience, taking listening pleasure to new dimensions predicated on the techné of the-body-in-virtual-space. Here I will suggest that TheWaveVR technology replaces the deployment of contemporary 'dance drugs' (such as MDMA, Ecstacy, LSD and amphetamines), which have been a prominent feature of dance music and social bonding since at least the 1960s (Tramacchi 2000). Moving into the late twentieth century, dance parties and dance club culture centered on this kind of drug-related experience (see Boys et al. 1997) and drug use was eventually woven into the very fabric of 'rave culture'. Des Tramacchi points out that:

> The term 'rave' was already in use during the 1960s to describe a 'psychedelic party' (Gore, 1997); however it has [now] become especially associated with a particular kind of gathering where people meet to 'dance for hours to music which is fast, loud and sounds like a machine's delight, while at the same time using drugs [and] gradually lose subjective belief in their self and merge into a collective body' (Jordan, 1995: 125). (Tramacchi 2000, 201)

Notably, there is a similar discourse deployed in discussions of TheWaveVR. Peter Rubin explains that "Throughout its young existence, TheWaveVR has worked with artists to create environments and interstitial shareable psychedelic experiences called 'Trips'" (2018, para. 6). In the case of TheWaveVR the listening body is produced through the interstices of techné and soma through which the bodymind orients itself within space: that is, the bodymind straddles mixed realities which provokes shifts in consciousness. For example, TheWaveVR hosted a listening party called N-FOLD, described as:

> a new synesthetic virtual reality experience, which pushes the boundaries of music, visuals, and social interaction. ... Known for his iconic visual work for artists such as The Weeknd, Erykah Badu, and Flying Lotus, David Wexler (Strangeloop) brings his trademark psychedelic aesthetic to the realm of VR, taking the user on a cinematic journey that could only exist using VR technology ... fans with an HTC Vive or Oculus Rift can be transported to the show to interact in a shared virtual space in real-time. (Mehow 2018)

Again, we see the capacity of this particular kind of technology to both build upon and extend previous frameworks of music listening and social bonding contingent upon the shared psychedelic experience. In a sense, TheWaveVR appropriates the pre-existing psychedelic aesthetic and even the psychedelic experiential mode in order to cultivate a zone that initiates and sustains a consciousness-shifting dance collective—without all the possible legal risks, social risks and health risks of 'dance club' drugs.

This is a significant shift for popular music listening practices in that the pursuit of psychedelic experience and the reach for shifts in consciousness are brought to bear through technicities which tend to circumnavigate the more problematic aspects of those encounters but retain the aspects from which listeners accumulate affective currency. For example, Tramacchi points to a kind of 'psychedelic morality,' of rave culture, in that many ravers note "the regular use of psychedelic drugs can be seen to promote morality and deep self-reflection" (2000, 209). These paradigms—of psychedelic experiences and consciousness expansion—have long been a concern in popular music listening, and TheWaveVR brings forth and reimagines these fundamental practices in divergent and imaginative ways. For instance, when a user enters TheWaveVR virtual dance club they are initiated into a space which is labelled The Social Lobby. According to TheWaveVR Reddit thread, The Social Lobby is "where the game really begins" because this is the space where the user can meet the other users and "try to make friends" (luter25 2017). It is from this point where the 'shareable psychedelic experience' begins to take shape:

> There is a lot to see in the lobby, wave deities, space dandelions, a cannon that looks surprisingly similar to a bong, and some mysterious floating currency called Samples. One of the key points to the social lobby ultimately, is to get prepped for events, and home parties, near the entrance there is a bar that has several orbs on display, all of which you can purchase with the samples you found while exploring. These boil down to two main categories, trips and toys. (luter25 2017)

The convergence of 'psychedelia' in shareable virtual space, replete with etiquette and currency, suggests a reimagining of Tramacchi's notion of 'psychedelic morality' whereby users interact in relation to a set of subcultural codes in order to construct and contribute to a safe and friendly environment to better explore the psychedelic music experience in a way that heightens its effect. In this case, the bodymind is produced by and through a technicity which, far from evading somatic encounter, builds upon its constituent parts in order to accentuate a listening experience in the virtual space.

The Somatechnics of Proprioception as Constituted in TheWaveVR

Much like the VR experimental art that Mark Hansen examines in *Bodies in Code* (2006), TheWaveVR enacts a *newly-coded* body for the listening subject, rather than a listening subject *sans* body. Hansen's work on theorizing VR artistic experiments provide a fitting point of departure to examine what kind of new or reimagined bodily experiences might be brought about by listening to music in the space of TheWaveVR. This aspect of Hansen's theoretical field is particularly concerned with what is often labelled as the sixth sense of bodily-beingness, that of proprioception (Hansen 2006, 26). Simon Gandevia and Uwe Proske tell us that, "proprioception includes the sense of position and movement of our limbs, the senses of muscle force and effort, and the sense of balance. These senses, triggered by our everyday activities, allow us to carry out our tasks successfully, without thinking" (2016, para. 2). This emphasis on the critical role of the body in VR, and its capacity for proprioception, is reinforced by Mark Mine et al. who explain that

> Manipulation in immersive virtual environments is difficult partly because users must do without the haptic contact with real objects they rely on in the real world to orient themselves and their manipulanda. To compensate for this lack, we propose exploiting the one real object every user has in a virtual environment, his [sic] body. (1997, 1)

With this emphasis in mind, I read the space of TheWaveVR as a reconstitution of the bodymind. That is, a bodymind produced by and through the technics of synesthetic and proprioceptive interactions with environment and Other, which is a genuine shift away from the prevailing ocularcentrism dominating music listening since at least the emergence of television in the 1950s and further accentuated during the 1980s rise of MTV listening culture. Much like Hansen's commentary on Char Davies' VR art experiments, this kind of technology "thus restores virtuality as a dimension of embodied life, as a technicity within the living rather than a (mere) technical artifact that affects life from the outside" (2006, 110). In this line of though, the listening subject—as they are produced by and with TheWaveVR—is reoriented in time, space and bodily-beingness in a way that reflects a kind of proprioceptive-centered experience or what Hansen would call "(one kind of) body-in-code" (110). In turn, this reimagines the very nature of listening to music as embedded in space, with others, and with the immaterial. In a true shift away from the emphases on material objects through which pleasure, in some ways, is contingent on haptic and tactile phenomena (such as the preoccupations with vinyl and so forth), TheWaveVR represents and constructs a redefined relationship between immateriality, sensuality, the senses, the bodymind, and listening pleasure.

However, what is also interesting about this redefined relationship is the ways in which previous imaginings of bodily-beingness are recapitulated in the virtual environment. For example, as explained on TheWaveVR Reddit thread, "Safety has been a recent focus in the application and the programmers have recently announced that a new blocking feature allows users to 'face their palm towards other users to block them'" (mushrooshi 2018). Much as in any material zone of social interaction, the 'precarious self' still lingers as a common concern in the potential to limit or impinge on listening pleasure in the shared space. This reinforces the importance and critical role of the users' body as it is experienced in the application, which is reminiscent of the ways in which Paul Verbeek thinks through the relationships between technologies and humans as a hybrid artifact of which "there is no pre-given subject and object" (Verbeek 2015, as cited in Yan Zheng 2017, 115). For Caroline Yan Zheng, "The boundaries between body, space, physical and digital have merged. Here, technology and materiality become one; coding and making interweave; the mediating power of computation couples with that from the aesthetics and sensual qualities of artefacts" (2017, 116). As articulated by one user on the experience of another, similar application:

> From the comfort of my couch in New Jersey I was able to virtually attend the intimate concert at the Belasco Theater in Los Angeles thanks to access via NextVR's app on Oculus. It was unlike any concert I ever attended or watched on TV because I experienced the concert from the stage. Rather than being a participant, I felt like a rock star. When Imagine Dragons lead singer Dan Reynolds belted out "I'm So Sorry," it was as if I was being pulled into the electrifying, emotional experience. ("The Next-Gen Concert" 2017)

In the 'new' scope of music listening the user is thoroughly invested in the bodymind, but in a way that is coded in and through the technicity of virtual mechanics.

Projection Mapping, Spatial Augmented Reality and Mixed Reality

However, not all encounters with virtuality require headsets. In this section, I focus on the emergent technologies which enable project mapping onto material space, also known as, 'spatial augmented reality,' as it deployed in the live concert arena. In this context, projection mapping deploys video projectors to map light onto a surface in order to render flat shapes into multi-dimensional models. Ramesh Raskar et al. describe the aim of Spatially Augmented Reality (SAR) as the creation of "an effective illusion of virtual objects coexisting with the real world" through rendering objects "directly within or on the user's physical space" (1998, 1). Though the first forays into SAR emerged in the late 1960s (John 2016; Bimber and Raskar 2005, 3), the public and commercial application of the technology is relatively new (Bimber and Raskar 2005, 7). Video projection, or 3D mapping does not require an application by the user but 'augmented reality' concerts require a smartphone with a correlate app, usually downloadable prior to the concert.

I examine both of these phenomena in relation to listening experience through the theoretical framework of 'mixed' or 'augmented' realities, whereby embeddedness becomes the central factor in the live/lived music experience and engagement is amplified tenfold. Mixed or 'hybrid' realities have become useful in discussions of this nature because:

> The notion of hybrid spaces assists in moving beyond unproductive binaries that position the virtual as simulation that exists in opposition to the real, to instead focus on the ongoing production of space, sociability, connectivity, flows, social relationships, and communication practices. (Grandinetti and Ecenbarger 2018, 3–4)

Thus, I read AR and 3D mapping in the live music experience as a technic that brings forth and extends the bodymind's connection to environment through constructions of embeddedness, and also as a technic that brings forth and extends the bodymind's connection to the musical experience, such as a bridge which connects the subject's inner and outer worlds. My contention is that the deployment of AR and video mapping functions to serve the imaginative and creative encounters that are already at work in the live music experience by borrowing from the bodymind's own somatic techné.

First, in relation to projection mapping, listening pleasure is built upon a multi-sensory synesthetic experience which stimulates the bodymind in a holistic encounter. Projection mapping calls upon the body's somatechnic through playfulness with 'presence.' According to Michaela French, "the convergence of light, body and technology can mediate states of presence … in immersive projection environments" (2017, 81). For example, Japanese band SID performs with a truly impressive set of visual designs produced through the video mapping technology (Putri Aziz 2013), in which the concert audience is ultimately immersed in an imaginative and creative field of experience. French indicates that, "Presence is elicited when overlapping modes of presentation and perception come together within a mediated space to construct an experience (Power 2008, 206), in which an illusion of non-mediation occurs" which provokes "experiences that build upon the fundamental human relationship with light to initiate moments of grace, in which the role of technology is eclipsed" (82). What is interesting about the construction of this experience, in relation to an understanding of somatechnics, is that this technology relies on converging the technics of both the bodymind and the mapping schema. That is to say that in order to achieve this effect, the function of 'first-order' presence (the sensorial interpretation of stimuli by the bodymind system) must integrate with the function of 'second order' presence, which is a "'a psychological state or subjective perception' (ISPR 2016) mediated by technology" in which "one is forced to perceive two separate environments simultaneously: the physical environment in which one is actually present, and the environment presented via the medium" (French 2017, 83). Thus, the resulting sense of presence, or what French describes as 'grace,' is produced through the technology of the projection mapping, which in fact, is simultaneously borrowing from the bodily technology. French unpacks this phenomenon by explaining that, "Presence is inseparably entwined with our embodied experience of light" (84) and that "we construct, narrate and manipulate spaces through the interaction of light and body, creating technologies that allow us to see beyond our embodied limits" (84). In the words of Elena del Río "technology springs from the very human condition of embodiment" (as cited in Sobchack 2004, 137). In the case of the band SID, which I note above, the individual is immersed within a magnificent visual architecture produced by the mapping technology. The music fan experiences the arrival of helicopters, the explosion of virtual pyrotechnics, the grace of enormous butterflies and a variety of surreal objects manifesting as presence in time and space. Live listening experience is utterly transformed by the adaptation of the mapping technology which, in many ways, produces new dimensionalities.

In perhaps an even more technical development of immersive music listening technologies, 'augmented reality' deployed in unison with the attendee's smartphone device provides another interesting point of departure in the 'new music' experiential category. I use the example of U2's recent 'augmented reality tour' to navigate this technology. In 'augmented reality' concerts, the attendee is asked to download a particular app onto their smartphone before the show, through which they can 'view' the visual augmentations to the concert space (in addition to U2, Eminem and Starset Juniper have both successfully offered this service). In the case of U2, a screen recording of a live concert performed in Tulsa demonstrates an opening scene overwhelmed with a magnificent waterfall emanating from the stage and pouring water over the crowd, with what appear to be star-filled, neon-green, serpent-like creatures shooting into the audience space (radiumsoup 2018). In an interview about the tour, Bono explains that the technology is about the traditional goals of concert experience, just like "stage-diving" was once "breaking the fourth wall" of the stage space, this kind of technology is a development in the reach for connection between artist, music, and attendee. For Bono, it is all about, "trying to *get* to our audience, trying to touch them, trying to reach them" (Fischer and Moorhead 2018). The discourse centered on connection resonates with the kind of theoretical framework in which I begin this chapter, in that the 'newness' of these technologies are contingent upon reworking and reimagining the potential of what music fans already love about music listening—that is, feelings of connection, pleasure, social bonding, and affective encounter. Augmented reality, as deployed in this case, also echoes Mark Hansen's assertions that, "the mixed reality paradigm" demonstrates new extensions and imaginings not of technology, but of the *body* (Hansen 2006, 5), a point which takes this argument full circle, back to the initial assertions I make regarding the centrality of the somatic in the virtual field of relations.

The Monstrous Holographic: Bringing Back the Dead

In the following section, I look at the experience of digitally projected performances which can be produced through a variety of technologies, such as holographic imagery or 2D projections of digital images on to metallicized surfaces to produce a three-dimensional effect. What concerns me here is not necessarily the variety of technologies used to produce such effects, but the kinds of emergent relationships which problematize listening pleasure in the live space, as a result of a somatechnicity that bridges Self and Other. This discussion also focuses exclusively on the relationship between once-living performers and fan listening culture, rather than non-referential digital agents such as Hatsune Miku (for a discussion of which see Connor 2016). This is because, as I will argue now, the once-living/now-dead binary produces a particular kind of problematic in the construction of listening pleasure.

Perhaps one of the most contentious encounters with the holographic 'musician' has been the performance of Tupac Shakur at the Coachella Festival in April of 2012. The performance was received in a variety of ways that ran the gamut from ecstatic appreciation to shock and awe (Martin 2012). However, it is the response of one music reporter which encapsulates an impact of the encounter on the listening process that I wish to focus on here. Laura Martin asserts that, "The Tupac hologram is macabre in a way that I cannot quite shake" (2012, para. 2). Here I will look at the way that the relationship between the listener (the Self) with the non-body of the dead artist (the Other) can, in some cases, deform listening pleasure by unsettling the performer/listener relationship.

Certainly, in many ways and for many fans, this kind of holographic technic "enriches and may even emancipate the experience of rock performance" (Arnold 2015, 178) but moreover, and perhaps more urgently, there are critical issues which are covered over by this technicity (see Little 2013). To begin with the Tupac 'hologram,' the listening experience is coproduced in the exchange of pleasure with the dead *non*-body. I echo Regina Arnold who elucidates that:

> The real Tupac Shakur was, like all humans, a complicated being. But Shakur's holographic body, dancing above the crowd at Coachella in its transparent trousers, is a cartoon thug, comely and compelling, hardly confrontational. Appearing as it does in the midst of a nostalgic revisitation of gangsta rap, his hologram can be read as merely a racially fetishized aspect of his humanity as a whole. (Martin 2012, 178)

Arnold is pointing to the ethics of race and the body here, particularly as it relates to the legacy of the African American male body as constituted by colonial violence. Arnold explains that, "One legacy of slavery is that images of black male bodies haunt American culture" and as just one example, she notes that "Billie Holiday famously sang about 'black bodies swinging in the Southern breeze,' in her song 'Strange Fruit' (1939)" (178). Arnold's reading is astute, in that this 'bringing back of the dead' is not, and cannot be, an apolitical or neutral act, especially as it is directed by the corporate entertainment system. For Jean Baudrillard, as Ken McLeod points out in relation to the Tupac 'hologram,' "the hologram is merely an extension of the hyper-real experience of music that has been the result of electronic recording and music reproduction practices over the past century" (2015, 118). I would concede that the hologram produces a site of pleasure contingent upon the existing framework of screen culture and ocularcentrism developed throughout the twentieth century certainly, regardless, that does not render the hologram unproblematic nor does it mean that holographic imagery is the *same* as other types of representation, such as the music video. I would argue that even though

Tupac's pre-existing TV images, merchandise, or rap videos that also (like holograms) might be said to haunt culture in a similar way, in fact differ in one crucial and critical way: Tupac himself, as an agentic black man, chose which songs to perform, with who, at which locales, and chose which videos to construct. He likely had some creative contribution as to how those videos might be constructed. Certainly, it can be argued that ultimate creative control is always-already compromised by the strictures of record company contractual obligations, especially in regard to merchandise and posthumously released albums. However, those limits sit a very long distance, theoretically and figuratively, from the activation of a dead man's ghost under the complete control of another individual or group, such as one would a marionette.

Perhaps one might even argue that, given the artist is dead, the agency of that individual is made redundant. I would counter this by noting that these are not concerns *for* the dead (although, ethically I do see this as an issue). Rather these are concerns about the ways in which the *culture* enables and allows corporate entertainment to change and rearrange Tupac's already existing memory and legacy. As Arnold writes poignantly, "inserting an iconic and politically-charged singer, whose work in the 1990s redefined rap as a site of counter-hegemonic speech, into a benign entertainment context, permanently and inevitably changes the symbolic weight of his oeuvre" (2015, 178). On this point, it is important for the reader to note that the holographic image was not simply a replaying of earlier footage that Tupac himself had already performed. Rather:

> The image [of Tupac] which was created by Digital Domain Media Group and projected by AV Images on a piece of mylar stretched across a mirror, was created pixel by pixel, frame by frame, from scratch. According to Ed Ulbrich, a member of the company the concert sequence was not simply a projection of already seen material. 'This is not found footage. This is not archival footage. This is an illusion.' (Arnold 2015, 180)

As such, the argument that the process commits a violent appropriation of his humanity and agency is even more compelling. On a technical note, again, producers can make the Tupac hologram 'do whatever they like,' resonating with the symbolic violence of African-American slavery still permeating U.S. culture. In a further theoretical maneuver on this problematic, John Freeman explains that, the holographic Tupac puts the finishing touches on what Darnell L. Moore describes as the dominant culture's 'ignorant escapist moves to disjoin the [Black] body from the spirit or mind;' more disturbing still, it furthers the project in which that Black body has been 'imaged, treated, and used as a marketable good.' Indulging in such spectacles continues what David Marriott describes as 'the spectacle of black lives consumed painlessly, virtually, teletechnologically... the commodity value of black death-in-life.' (2016, para. 5)

However, the question, for *this* discussion, is how this technicity plays upon the politics and pleasure of listening. I argue that this kind of 'marionetting' produces a power of horror through which the listener must reorganize notions of the body, death and the technology of resurrection, resulting in what Lauren Martin describes above as a sense of the 'macabre that one cannot shake.'

To approach this in theoretical terms, the hologram—particularly that of the black male body that is, and has been, the site of brutal appropriation—is thus representational of the non-body of the marginalized Other. The specter of the colonial violence which refuses to sink back into non-memory confronts the listener in the present—with all the horrors of the past. This can be read through Kristeva's notion of abjection, that is, the subject's revulsion to that which threatens to collapse all meaning (1982, 3). For Kristeva, the ultimate provocation of this process is the corpse, which points to the subject's materiality, and ultimate death (3–4). In resonance of this reading of the black holographic body, in *Powers of Horror* Kristeva reminds us of the Borges' fable regarding "The Cruel Redeemer Lazarus Morell" in which, "the frightful redeemer raises his slaves from the dead only to have them die more fully, but not until they have been circulated—and have brought in a return—like currency" (1982, 24) In this regard, what is Tupac's hologram but an attempt, both produced by light and veiled by that very technicity, to resurrect the labor value of the black male body?

Thus, one imagines the sense of death and the dead body as integrative and integrated into the experience of the performance, ultimately leading to an ambivalence brought forth by somatechnics of other-self relations and mediated by extant technological sites. Kristeva might say, in this case, that:

There looms, within abjection, one of those violent, dark revolts of being, directed against a threat that seems to emanate from an exorbitant outside or inside, ejected beyond the scope of the possible, the tolerable, the thinkable. It lies there, quite close, but it cannot be assimilated. It beseeches,

> worries, and fascinates desire, which, nevertheless, does not let itself be seduced. Apprehensive, desire turns aside; sickened, it rejects. A certainty protects it from the shameful—a certainty of which it is proud holds on to it. But simultaneously, just the same, that impetus, that spasm, that leap is drawn toward an elsewhere as tempting as it is condemned. Unflaggingly, like an inescapable boomerang, a vortex of summons and repulsion places the one haunted by it literally beside himself. (1982, 1)

Following this, I am proposing here that the pleasure and politics of this particular listening experience is 'summoned into this vortex of repulsion' and one is 'haunted' quite literally by the holographic non-body. This kind of technological mediation brings forth, in autopoietic eddies and currents, an ambivalence marked by the kinds of descriptions we see in public discussions of said practice, those of the macabre and shock and awe, as noted above. One might say, in the case of Tupac, the body of the Other is unhinged from its humanity and thus challenges the pleasure, as well as the ethics, of listening.

Conclusion

What culture considers to be 'new' is variable and contingent upon a variety of meaning-making mechanisms which construct the ways in which we understand technology to be meaningful in any given context. For example, in mainstream discussion, 3D film is often referred to as 'new' and its emergence is often chronicled in relation to James Cameron's 2009 blockbuster *Avatar*. However, 3D film was popularised, at least in the U.S., in the mid-twentieth century, but without enough commercial traction, simply 'went away' until it re-emerged in the 2000s (Christie 2014). Lisa Gitelman remarks that "all media were once new" but it is by examining "the novelty years, transitional states, and identity crises of different media" that we can learn so much more "about the course of media history and about the broad conditions by which media and communication are and have been shaped" (2006, 1). Similarly, I examine the emergence of virtual and mixed reality technologies in the context of ubiquitous digital media devices and the discursive practices, often emerging through sci-fi film and novels, which have produced notions of virtual technology as 'futuristic.'

Moreover, in this chapter, I have suggested that certain shifts brought about by the introduction and uptake of VR and MR technologies have produced a discourse of 'newness' that actually tells us something about the status of the bodymind. This discourse tells us that the bodymind is the central focus in new technologies, despite previous anxieties about the potential dissolution or collapse of the flesh into the cybernetic fold. I have argued, following Mark Hansen, that VR and MR consummates the insistence of the *totality* of sensory faculties as the primary basis of all listening experience—from the proprioceptive faculty to visual functions which reshape social consciousness and psychedelic traditions. Bridging this argument from Chap. 2, in which I suggested that listening experience became bound, in very specific and unique ways, to sensory phenomena—what I called a *sensory somatechnics*—this chapter rounds off that examination to suggest that the future of listening culture lies in the playful interactions between corporeality and virtual techné.

References

Apatow, Judd, Lou Holtz, Jr., and Ben Stiller. 1996. *The Cable Guy*. Directed by Ben Stiller. Los Angeles, CA: Columbia Pictures.

Arnold, Regina. 2015. There's a Spectre Haunting Hip Hop: Tupac at Coachella, Holograms in Concert, and the Future of Live Performance. In *Death and the Rockstar*, ed. Catherine Strong and Barbara Lebrun, 177–187. London: Ashgate.

Attali, Jacques. 1985. *Noise: The Political Economy of Music*. Minneapolis: University of Minnesota Press.

Bimber, Oliver, and Ramesh Raskar. 2005. *Spatial Augmented Reality: Merging Real and Virtual Worlds*. Wellesley, MA: A.K. Peters.

Boys, Annabel, Simon Lenton, and Kath Norcross. 1997. Polydrug Use at Raves by a Western Australian Sample. *Drug and Alcohol Review* 16 (3): 227–234.

Christie, Ian. 2014. Will the 3D Revolution Happen? Will the 3D Revolution Happen? A Brief Perspective on the Long History of Stereoscopy (with Special Thanks to Eisenstein and Bazin). In *Teché/Technology: Researching Cinema and Media Technologies, Development Use and Impact*, ed. Annie van den Oever, 115–135. Amsterdam: Amsterdam University Press.

Conner, Thomas. 2016. Hatsune Miku, 2.0Pac, and Beyond: Rewinding and Fast-Forwarding the Virtual Pop Star. In *The Oxford Handbook of Music and Virtuality*, ed. Sheila Whitely and Shara Rambarran, 129–147. New York: Oxford University Press. https://williamlittlesociology.wordpress.com/2013/02/23/tupac-is-a-hologram-the-re-history-of-rocknroll/.

Fischer, Lisa, and Jeremy Moorhead. 2018. Bono and the Edge: Why U2 is Embracing AR Tech on Tour. *CNN*, May 2. https://money.cnn.com/video/technology/2018/05/02/u2-concert-bono-edge-augmented-reality-laurie-segall-orig.cnnmoney/index.html

Flew, Terry. 2014. *New Media*. South Melbourne: Oxford University Press.

Freeman, John. 2016. Tupac's 'Holographic Resurrection:' Corporate Takeover or Rage against the Machinic? *ctheory.net*, April 6. http://ctheory.net/ctheory_wp/tupacs-holographic-resurrection-corporate-takeover-or-rage-against-the-machinic/#_ednref8

French, Michaela. 2017. Bodies in Light: Mediating States of Presence. In *Digital Bodies: Creativity and Technology in the Arts and Humanities*, ed. Susan Broadhurst and Sara Price, 81–95. London: Palgrave Macmillan.

G-Dub. 2018. How To: Making Friends in TheWaveVR. *YouTube Video*, July 22. https://www.youtube.com/watch?v=9bDFdG7-RGo&ab_channel=G-Dub

Gandevia, Simon, and Uwe Proske. 2016. Proprioception: The Sense Within. *The Scientist*, September 1. https://www.the-scientist.com/features/proprioception-the-sense-within-32940

Gitelman, Lisa. 2006. *Always Already New: Media, History, and the Data of Culture*. Cambridge, MA: MIT Press.

Grandinetti, Justin, and Charles Ecenbarger. 2018. Imagine Pokémon in the "Real" World: A Deleuzian Approach to Pokémon GO and Augmented Reality. *Critical Studies in Media Communication* 35 (5): 440–454.

Hansen, Mark. 2006. *Bodies in Code*. New York: Routledge.

Hesmondhalgh, David. 2013. *Why Music Matters*. Chichester: Wiley-Blackwell.

John, Brandon. 2016. How A/V Sets are Changing the Future of Live Music. *Tone Deaf*, November 14. https://tonedeaf.com.au/av-sets-changing-the-future-of-live-music/

Kristeva, Julia. 1982. *Powers of Horror*. New York: Columbia University Press.

Little, William. 2013. Tupac is a Hologram: The Re-History of Rock'n'Roll. *William Little's Sociology Web Pages*, February 23.

luter25. 2017. Welcome to TheWaveVR. *Reddit.com*. https://www.reddit.com/r/TheWaveVr/comments/66ytzo/welcome_to_the_wave_vr/https://www.meetup.com/JumpIntoTheLightVR/events/240341012/

Marcus, Ezra. 2016. 'TheWaveVR' is the Virtual Reality Concert Platform of the Future. *Thump*, September 30. https://thump.vice.com/en_ca/article/ez7mdp/wave-virtual-reality-future

Martin, Lauren. 2012. The 2.0Pac Hologram was Really, Really Weird. *Dummy Mag*, April 20. https://www.dummymag.com/features/the-20pac-hologram-was-really-really-weird

McLeod, Ken. 2015. Hip Hop Holograms: Tupac Shakur, Technological Immortality and Time Travel. In *Afrofuturism 2.0: The Rise of Astro-blackness*, ed. Reynaldo Anderson and Charles E. Jones, 109–124. London: Lexington.

Mehow. 2018. Take a Trip in TheWaveVR; Special Event N-Fold Release Party. *Jump into the Light, Meetup.com*, June 28.

Mine, Mark, Frederick Brooks, Jr., and Carlo Sequin. 1997. Moving Objects in Space: Exploiting Proprioception in Virtual-Environment Interaction. In *Proceedings of the 24th Annual Conference on Computer Graphics and Interactive Techniques*, 19–26. New York: Addison-Wesley Publishing.

mushrooshi. 2018. TheWaveVR 0.11.0: Player Safety Features, Maintenance and Optimization. *Reddit.com*. https://old.reddit.com/r/Vive/comments/93g3ap/thewavevr_0110_player_safety_features/?st=jn6wopi3&sh=c9f46c9e

Putri Aziz. 2013. Mapping SID Concert Japan. *YouTube*. https://www.youtube.com/watch?v=h_2fT9lRWK0&feature=youtu.be

radiumsoup. 2018. U2 E&I in Concert AR Effect Experience Tour Opener Tulsa, 2 May 2018. *YouTube*. https://www.youtube.com/watch?v=tOUtd4UKs78&feature=youtu.be

Raskar, Ramesh, Greg Welch, and Henry Fuchs. 1998. Spatially Augmented Reality. *First International Workshop on Augmented Reality*, 1–7, San Francisco, CA, November 1.

Rubin, Peter. 2018. You Can't See Ready Player One Yet, but with VR You Can Dance in It. *Wired*, March 13. https://www.wired.com/story/ready-player-one-vr/

Sobchack, Vivian Carol. 2004. *Carnal Thoughts: Embodiment and Moving Image Culture*. Berkeley: University of California Press.

The Next-Gen Concert Experience: Imagine Dragons in Virtual Reality. 2017. *Wired Magazine*. https://www.wired.com/brandlab/2017/07/next-gen-concert-experience-imagine-dragons-virtual-reality/

Tramacchi, Des. 2000. Field Tripping: Psychedelic Communitas and Ritual in the Australian Bush. *Journal of Contemporary Religion* 15 (2): 201–213.

travtrigs. 2018. I'll be Spinning Glitch Hop & Dubstep in TheWaveVR Tonight @ 8:30 PT. *Reddit.com*. https://www.reddit.com/comments/99t2jy

Wiley, S.B.C. 2005. Spatial Materialism: Grossberg's Deleuzean Cultural Studies. *Cultural Studies* 19 (1): 63–99.

Yan Zheng, Caroline. 2017. Machinizing Humans and Humanizing Machines: Emotional Relationships Mediated by Technology and Material Experience. In *Digital Bodies: Creativity and Technology in the Arts and Humanities*, ed. Susan Broadhurst and Sara Price, 111–128. London: Palgrave Macmillan.

CHAPTER 7

Conclusion: Music as Somatechné

Introduction

Bodies are the site of continuing political and social tension. The body, as an object of culture, is monitored, surveilled, often pulled apart, sometimes reassembled, and always-already produced by techné. The way the body (or as I have referred to it in this book—the bodymind) comes to know listening pleasure is a complex process and subject to the status of that bodymind in any given context. I have traversed the relationships between listening pleasure, historical context and technicity in a way that offers some reflection on the intersecting and irretrievable entanglements between this triad. To do so has led me to one final conclusion: That *music itself* is a technicity which produces and constitutes the lived formation in some of the most unique and thoughtful ways. What I mean by this assertion is that, at the root of all the technologies through which music is mediated, it is the primary aspects of music which provide us an ultimate technical gift. From the organization of sounds to the tempo of beats, to the granularity of vocality and the all the details in between—it is music which we might call an exteriorization in the pursuit of pleasure and, as this book has detailed, music culture frames listening as one of the most significant human pursuits. To end this book, I will explore the notion of music itself as somatechné.

L. Glitsos, *Somatechnics and Popular Music in Digital Contexts*, Pop Music, Culture and Identity,
https://doi.org/10.1007/978-3-030-18122-2_7

Music as Somatechné

The body brings music into being. Music is most aptly defined, by John Blacking, as "humanly organized sound" (Blacking 1973, 10). The connection between corporeality and the phenomena of musical texts is ultimate, and their position to each other is simultaneously produced; the movement of the body "occupies a singularly central position in the production of music" (Mason 2012, 6). This is a technicity in and of itself, as Paul Mason puts it, "The conscious manipulation of body and sound through time and space depends upon the procurement of skill through individual, social and cultural development" (6). In the most integral way, music is techné which emerges from the deepest wells of the bodymind. This is what Jean Luc Nancy might term, "ecotechnics," which is otherwise to say that, "'Creation' is the techné of bodies. Our world creates the great number of bodies, creates itself as the world of bodies" (89). Music presents us with an exemplary model of the ecotechnical—the relation between exteriorization and interiorization.

However, simultaneously, *music brings the body into being*. How we understand the body to be, and what the body is capable of, can be wrought through the interconnection of the bodymind and the musical text, or the 'humanly organized sound'. If technicity produces the body in the world, then, as a deeply complex techné of organized sound, music produces a very specific kind of body in the world at any given point. Music invents the body anew. Alf Gabrielsson's insightful ethnographic research reveals the depth and extent of the ways in which we can understand music as producing the bodymind 'anew'. Gabrielsson's research focuses on exploring 'strong experiences with music'. In many ways, he reports on musical experiences which I would describe as somatechniques, which grant access to a kind of inner emotional delta. For example, in one response from Gabrielsson's research, during a concert of Beethoven's ninth symphony, a participant reports: "I felt calm and happy, and in particular during the slow third movement I felt a wave of almost extraterrestrial joy go through me, more or less the same feeling as being head over heels in love" (2011, 110). Then, in a suggestion of a physical purge, the participant adds that, "When I closed my eyes, the tears started to fall. The music somehow flooded directly into me, and my senses were wide open" (110). In this case, music is a permeating force which functions as a very specific kind of emotional agent for this individual, or in other words, it functions as a somatechnicity brought about through the complementary techniques occurring in both music and bodymind. Metaphorically, we could say that the music functions like the technology of a key, while the bodymind functions as a locking mechanism. In some senses, music 'unlocks' the bodymind.

Music is also a technology through which consciousness can be modified and altered. Following on from the same survey response above, the participant provided insight into a kind of *merging*, of bodymind with music, stating that after this particular show, "We met some of our friends who stood chatting, but I wasn't really present. The music was still inside me, and I was inside the music" (111). Far from being a unique anecdote, Gabrielsson's research provides several of these accounts. For instance, another survey participant finds that in very strong experiences with music, "In some way, time and space disappear. You sit there breathless and listen and think that you are a part of the music in some way" (2011, 113). Yet again, in another survey response within Gabrielsson's research, a listener claims that they were listening to David Oistrakh (a classical violinist) and underwent some sort of shift in consciousness. The participant recounts that:

> I was on my own, it was half-dark in the room, I lay on my bed and was totally unprepared for what was going to be played. All this, I think, contributed to my mood being the right one for the experience with the music to be so strong. I was completely numbed by the music. I forgot time and space. My body vanished from my consciousness. I was totally immersed in the music. (Gabrielsson 2011, 111)

This is what is meant by stating that music and the body produce each other simultaneously, it is a somatechnical relationship in that the music/bodymind relationship offers a "dynamic means in and through which corporealities are crafted, that is, continuously engendered in relation to others and to a world" (Sullivan and Murray 2009, 3). In these instances, the participants are reporting a collapse between the two territories (music and subjectivity), which is a becoming, or a changing. The process is dynamic because each artefact plays off and upon the other. The musical text is a product of human dynamism, but so too, human dynamism calls upon music in order to be moved.

Finally, I suggest music as a somatechnique through which discomfort or pain might be relieved. In an account from a woman, we find the experience of music listening permeates the very bone:

> In the old days, in my childhood, [the 1910s] they didn't usually do any fillings on milk teeth. Anyhow, whatever the reason, I often had a toothache. My father came home one day with a gramophone with an enormous horn that I would have liked to crawl into. He had bought a lot of gramophone records too, mainly classical music. But also one or two lighter classics such as H. C. Lumbye's 'Champagne Galop.' When my toothache was at its worst, I wound up the gramophone and played the 'Champagne Galop.' Strange though it may seem, the pain went away. Every time! (Gabrielsson 2011, 262–263)

Again, far from being a unique account, Gabrielsson archives similar anecdotes in a range of contexts. Another woman found herself "mortally ill with fungus poisoning" but was unaware that she ingested a toxic mushroom. She recalls that:

> Later, at a mushroom exhibition, I found out that this mushroom is now regarded as decidedly poisonous. Anyhow, I had a very painful night, vomiting and having cramps. I had the radio turned on. At about five in the morning, I was suddenly struck by the thought that my heart might not manage this. Just then, there came a piece by Mozart on the radio; it was delightful. It was God and heaven and all angels, and I felt that now I had reached a turning point. And to the strains of that music I conquered the poisoning. (2011, 263)

The exact nature and influence of Mozart in this instance is ambiguous, but the argument rests on the assertion that this listener found their pain to be minimized and their will strengthened by music alone, acting in the same way as any other technology of medicine.

In fact, it is well known that for hundreds of years Amazonian shamans have been utilizing music in tandem with Ayahuasca as a medical aid to invoke physical purging and emotional release. Ayahuasca is a brew made from of the vine of the Ayahuasca plant, and usually mixed with another plant containing the active constituent of dimethyltryptamine, such as Chakruna. The word 'Ayahuasca' comes from the indigenous Andean language of Quechua, and is often translated as "vine of the soul" or "vine of the dead" (Kjellgren et al. 2009, 309). The Ayahuasca vine is endemic to the upper Amazonian region and is often reported to invoke "powerful visions [...] personal insights, intellectual ideations [...] and profound spiritual and mystical experiences" for some individuals (Shanon 2002, 13). The vine is used as a sacrament in select recognized religions, such as the Brazilian church of Santo Daime (Gorman 2010, 8). In both the Church sacrament and the indigenous context, the music is an absolutely integral aspect in the Ayahuasca experience and functions as a kind of medico-spiritual guide for the participant as they journey through shifts in consciousness (see Narby 1998; see Shanon 2002). Elsewhere, I have noted that this tradition has re-emerged in contemporary contexts with renewed interest, particularly in the Indie genre of popular music, in which forms of pre-modern ecotechnics have been translated for the digital context (Glitsos 2014). Moreover, these examples demonstrate the way music functions as that which brings the body into the world anew, that is, with reorganized affective architectures. Nancy tells us that:

> The ecotechnical functions with technical apparatuses, to which our every part is connected. But what it makes are our bodies, which it brings into the world and links to the system, thereby creating our bodies as more visible, more proliferating, more polymorphic, more compressed, more 'amassed' and 'zoned' than ever before. Through the creation of bodies, the ecotechnical has the sense that we vainly seek in the remains of the sky or the spirit. (Nancy 2008, 89)

Music, similarly, crafts bodies as polymorphic, that is, fluid and changing as they are exacted upon by phenomenal forces. Perhaps, as above, music might generate a bodymind that experiences new insights brought about through shifts in consciousness, or music may negotiate the very pain centres of the somatic. In a variety of ways, music is somatechné by way of a crafting through the intersections of interiority and exteriorization, the 'horizon of possibility' as Stielger may suggest.

Concluding Remarks

In many ways, this book describes and frames a listening culture that is transitioning from building connections through traditional modes of listening associated with materiality born from twentieth century contexts. The transition marks a shift to building connections through sharing digital products on new media platforms. This transition is messy, non-linear, and at times, contradictory. For example, following Bartmanski and Woodward (2015), I have noted the continued preoccupation with the material objectification of musical texts and how this preoccupation comes to invoke new meanings in listening culture in the context of digitization. Different modes of consumption can and do exist simultaneously, however, they also produce conflicting emotions and destabilizing effects in their differences.

Framing the bodymind in context also means bearing down on the implications of the fast and sometimes even radical shifts culture experiences as the result of digitization and the ubiquity of digital devices. The post-millennium digital context is still awash with the previous two decades of discursive practices building the cultural significance and meaning of Web 2.0. Tim O'Reilly and John Battelle coined the term 'Web 2.0' at a Summit on Internet technology in 2004 (O'Reilly and Battelle 2009, 1) referring to the second wave of technical features developed for Internet technologies, such as the adoption of "public application programming interfaces" that enable communication between users and the embedding of rich media such as video (Cormode and Krishnamurty 2008). As a result of these new technical features, Web 2.0 has cultivated a "participatory culture" (Jenkins 2006), "user-generated material" (Van Dijck 2009), exciting new models of "co-creation" (Choi and Burnes 2013), and a forum for transnational "crowd sourcing" (Howe 2006). Web 2.0 is the technology of our time, which hosts complex multi-levelled dimensions driven by affective phenomena that rescripts traditional notions of bodies, spaces and social bonding.

Even though Web 2.0 builds a virtual space, it produces contexts which reimagine the organic body because the virtual confuses and interrogates the limits of postmodern corporeality. Melissa Gregg and Gregory Seigworth explain that affect occurs in those encounters between bodies, "whether those bodies are defined as fully human, part-human, non-human or otherwise" (2010, 2). This approach can be read as an extension of Spinoza's concern with the relationships between bodies which, as Jenny Sundén explains "can be human bodies, but also body parts, nonhuman animals, and inanimate objects" (2013, 372). This point in particular is critical because I have suggested that the borders between what is/is not the body are difficult to define. For example, in the computer-mediated contexts of music listening, the computer interface and all its associated technologies colour the affective dimensions of listening in ways that both trace and confuse normative bodily boundaries. Technologies are just as much a part of the music they mediate as they are a part of the corporeal body entangled in the encounter. In the late 1980s, Donna Haraway reimagined a new body, predicated on the visions of the cyborgian subject and articulated by the language of cybernetics, writing: “We are all chimeras, theorised and fabricated hybrids of machine and organism; in short we are cyborgs” (1987, 2). Far from being vessels of empty robotic scripts, the cyborg represents an expanded subject, one that is liberated, if only partially, from many of the traditional bodily affects regulated by the discourses of biopolitics. For instance, we can read the configuration of the MP3 in the ways in which it parallels the cybernetic subject—both imply fluidity, mobility, and exist in their capacity to be replicated and shared. In the cyberspace, the subject shares data packets as forms of the self—from MP3s to YouTube clips. In the cybernetic fold, the music listening experience is radically redefined in line with the transformations of the technosocial bodymind.

The online experience plays upon the material body through reflexive pathways and thus *becomes the inhered into that bodymind*, especially through the screen/eye relationship. In Nigel Clark’s discussion of William Gibson’s cult cyberpunk fiction *Neuromancer* (1984), Clark writes that cyberspace can enact an “even greater intensity than the built environment” because “human bodies act as the receptive surfaces for the images projected by the media” (1995, 123). There is a relationship between the screen and the eye that creates imagined worlds for each individual, which can be just as elaborate, or even more elaborate, than the material world.

I began this book with a focus on the ways in which sensory perception comes to render music as both knowable and pleasurable through the interplay of bodymind with materiality—that is, through the complementary interactions with surface, weight, touch, feel, smell, size, and so forth. The changing notions around corporeality, knowability, perception, and materialism come to reshape our relationships to the physical objects of music, and this then shapes listening pleasure more broadly.

I have also elucidated on the various ways in which we might imagine new distinctions (or lack thereof) between liveness and mediatization in the post-millennium digital context by applying theoretical perspectives of somatechnics to certain live music listening practices. There are certain technics which bring forth and produce listening pleasure, functioning with and within the somatic, such as the technics of time and its significance in the production of liveness and authenticity; the somatechnics of dance as it functions to embody music listening; and the somatechnics of vocality as it functions in the live music space as a vibratory technology. The widespread uptake of mobile devices has also infiltrated the concert space and brought new understandings to bear on what it means to find pleasure in live music.

I have also been concerned with the way mobile music devices with touchscreen technology produce new somatechnical figurations that reshape emotional dynamics of music listening. Touchscreens imply the relationship between skin on skin—the skin of our body (in particular the hands) against the skin of the screen. The coming together of bodies, fluids and other organic materials which 'stick' to the touchscreen, the language of 'stickiness' pointed to Sara Ahmed's conceptualization of the way affect can "stick" to bodies (2004) and by "thinking through the skin," to use Ahmed and Stacey's words, I suggested haptic touchscreen technology is an exciting new way to imagine music listening in terms of cyborgian relations (Ahmed and Stacey 2003).

However, across the field of popular music studies, I do feel that fixed-point personal computers require more attention as portals for musical enjoyment. Thus, I focused on the Internet as a technology of music listening as it is accessed *exclusively* through fixed-point personal computers. We find here that the listener actively and creatively produces their own listening experience. In this context, the personal computer produces a unified field of relations, an artifact which I examined through the lens of somatechnics. This unified field of relations realigns and reconstitutes listening pleasure in unexpected ways because it enables unique and creative listening practices contingent upon the functional nature of Internet-computing technologies and they are expressed in the interactive Web 2.0 culture. This chapter argues that the personal computer bodymind produces a new 'type' of listener: the creative listener.

Finally, I have reflected on several practices of music listening which have emerged in the past decade and which integrate virtual or mixed realities such as TheWaveVR and 3D mapping at live shows. By doing so, I have proposed that popular music culture is seeing an emerging trend toward virtual and mixed reality technologies in music listening which play upon notions of 'the real' and a general concern with 'immersion'. Most importantly, this work expounds on what this concern might tell us about the bodymind in the music listening experience and how this experience continues to be produced by techné. The chapter offers that next-generation virtual reality and 'mixed reality' technologies reconstitute the somatechnic capacities of bodymind dynamics in listening practices in a way that foregrounds the site of the body as the 'ground zero' of all experiential exchange.

Altogether, this research functions to complement several other works emerging in the field of popular music studies which focus on the interconnections of bodies, technologies and music. Raphaël Nowak's text *Consuming Music in the Digital Age* (2016), which has provided insight into the "material modalities of music consumption" as they constitute life narratives and identity-making mechanisms, is particularly insightful. Further, Bartmanski and Woodward's examination on the continuing significance of vinyl in digital contexts has helped launch and stabilize many arguments in this book (2015). Ben Green's (2017) discussion of Wu-Tang Clan's unique compact disc release has also compounded the importance of these debates regarding the ongoing significance of materiality. These arguments have nicely complemented the style of 'sensorial somatechnics' which informs this work, particularly as laid out in Chap. 2. Perhaps most recently I have found Nick Prior's *Popular Music Digital Technology and Society* (2018) as an indispensable addition to the study of popular music and digital technology. Though Prior does not employ the language of somatechnics, the book demonstrates important entanglements as each emerge together, as one. However, there is still more work that can be done and if, as Attali claims, music is a herald of the future, then these findings suggest that while listening practices will evolve in complexity, they will continue to do so in deeply collective ways. For instance, spatial audio in live contexts is emerging as a vital force in the field of music production, and as Robert Lawrence remarks, "Many academics, researchers, audiophiles and producers consider the field of spatial audio as one of the final frontiers for high quality sound reproduction and recorded sonic immersion" (2019, 134). There are always new horizons through which we might reimagine listening pleasure for the popular music fan.

At the start of this book, I suggested this work to be a celebration of the lived formation as it is oriented by the experience of music listening. I hope that by reading through some or all of its contents, it has provided the reader with glimpses into the ways in which the bodymind is brought into the world through music.

References

Ahmed, Sara. 2004. *The Cultural Politics of Emotion*. Edinburgh, Scotland: Edinburgh University Press.

Ahmed, Sara, and Jackie Stacey. 2003. *Thinking Through the Skin*. Hoboken, NJ: Taylor & Francis.

Bartmanski, Dominik, and Ian Woodward. 2015. The Vinyl: The Analogue Medium in the Age of Digital Reproduction. *Journal of Consumer Culture* 15 (1): 3–27.

Blacking, John. 1973. *How Musical is Man*. Seattle, WA: University of Washington Press.

Choi, Hwanho, and Bernard Burnes. 2013. The Internet and Value Co-creation: The Case of the Popular Music Industry. *Prometheus* 31 (1): 35–53.

Clark, Nigel. 1995. Rear-view Mirrorshades: The Recursive Generation of the Cyber Body. In *Cyberspace/Cyberbodies/Cyberpunk: Cultures of Technological Embodiment*, ed. Mike Featherstone and Roger Burrows, 113–134. London: Sage.

Cormode, Graham, and Balachander Krishnamurty. 2008. Key Differences Between Web 1.0 and Web 2.0. *First Monday* 13 (6).

Gabrielsson, Alf. 2011. *Strong Experiences with Music: Music is Much More Than Just Music*. Oxford: Oxford University Press.

Gibson, William. 1984. *Neuromancer*. New York: Ace Books.

Glitsos, Laura. 2014. The Ecology of 'Sacred Space:' Indie Music's Exploration and Construction of Sacred Space in the Context of Contemporary Digital Music. In *Communities, Places, Ecologies: Proceedings of the 2013 IASPM-ANZ Conference*, ed. Jadey O'Regan and Toby Wren, 138–146. Brisbane: International Association for the Study of Popular Music.

Gorman, P. 2010. *Ayahuasca in My Blood: 25 Years of Medicine Dreaming*. Texas: Peter Gorman Bench Press.

Green, Ben. 2017. Having the Sceptre: Wu-Tang Clan and the Aura of Music in the Age of Digital Reproduction. *Popular Music* 36 (3): 427–440.

Gregg, Melissa, and Gregory Seigworth. 2010. *The Affect Theory Reader*. Durham, NC: Duke University Press.

Haraway, Donna. 1987. A Manifesto for Cyborgs: Science, Technology, and Socialist Feminism in the 1980s. *Australian Feminist Studies* 2 (4): 1–42.

Howe, Jeff. 2006. The Rise of Crowdsourcing. *Wired Magazine*, June 14. http://www.wired.com/wired/archive/14.06/crowds_pr.html

Jenkins, Henry. 2006. *Fans, Bloggers, and Gamers: Exploring Participatory Culture*. New York: New York University Press.

Kjellgren, A., A. Eriksson, and T. Norlander. 2009. Experiences of Encounters with Ayahuasca: 'The Vine of the Soul'. *Journal of Psychoactive Drugs* 41 (1): 309–315.

Lawrence, Robert. 2019. Mixing Music for Immersive Audio Experiences. In *Producing Music*, ed. R. Hepworth-Sawyer, J. Hodgson, and M. Marrington. London: Routledge.

Mason, Paul. 2012. Music, Dance and the Total Art Work: Choreomusicology in Theory and Practice. *Research in Dance Education* 13 (1): 5–24.

Nancy, Jean-Luc. 2008. *Corpus*. New York: Fordham.

Narby, Jeremy. 1998. *The Cosmic Serpent: DNA and the Origins of Knowledge*. New York: Penguin.

Nowak, Raphaël. 2016. *Consuming Music in the Digital Age: Technologies, Roles and Everyday Life*. Hampshire: Palgrave Macmillan.

O'Reilly, Tim, and John Battelle. 2009. *Web Squared: Web 2.0 Five Years on: Web 2.0 Summit 2011*. San Francisco, CA: O'Reilly Media.

Prior, Nick. 2018. *Popular Music, Digital Technology and Society*. London: Sage.

Shanon, Benny. 2002. *The Antipodes of the Mind: Charting the Phenomenology of the Ayahuasca Experience*. Oxford: Oxford University Press.

Sullivan, Nikki, and Samantha Murray, eds. 2009. *Somatechnics: Queering the Technologisation of Bodies*. Surrey: Ashgate.

Sundén, Jenny. 2013. Corporeal Anachronisms: Notes on Affect, Relationality, and Power in Steampunk. *Somatechnics* 3 (2): 369–386.

Van Dijck, José. 2009. Users Like You? Theorizing Agency in User-Generated Content. *Media, Culture and Society* 31 (1): 41–58.

Index[1]

[1] Note: Page numbers followed by 'n' refer to notes.

L. Glitsos, *Somatechnics and Popular Music in Digital Contexts*, Pop Music, Culture and Identity,
https://doi.org/10.1007/978-3-030-18122-2

Kolibri 360 GmbH
Ferdinand-Jühlke-Straße 7
99095 Erfurt, Deutschland
produktsicherheit@kolibri360.de